建筑立场系列丛书 No.5

C3

# 小型建筑

# SMALL Variations

韩国C3出版公社 | 编

王单单 李群 赵珊珊 张媛媛 李硕 高圆 曲艺 闫文诗 赵敏 | 译

大连理工大学出版社

4

# 小型建筑的变异

## 单一型、复杂型和雕塑型



84

住宅设计

# 住宅和城市景观



158

# BIG未来之窗

## 4 SMALL Variations

### Single, Sophisticated and Sculptural



## 84 DWELL HOW Housing and Urban Landscape



## 158 BIG Window to the Future

# 小型建筑的变异
# 单一型、复杂型和雕塑型
# SMALL Variations
## Single, Sophisticated and Sculptural

1995年，雷姆·库哈斯以“小，中，大，超大”四个词汇提出了一种建筑研究的新方法。这四个词汇改变了评论家们的评论方式，并为针对建筑体量的分析提出来了新的线索。他提出的两种极端情况——小型和超大型，极大地影响了近年来的项目规划。

在接下来的章节中，我们将进一步细化对小型建筑的研究，并将其分为三类：单一型、复杂型和雕塑型。围绕“小”的概念，金基德导演的电影、马克·安吉利的作品、卡罗·莫里诺的摄影以及博格斯的疯狂系列作品都可以帮我们找出“小”规模分析的连贯性，在微观世界中完成复杂的行为活动。

In 1995, Rem Koolhaas fixed the coordinates of the architectural survey around four words, S, M, L, XL, changing the trajectory of the critic, which will find in the size of the buildings new cues for analysis. The two extremes of the proposal, small and extra-large, in particular, have significantly influenced the projects of recent years.
In the text that follows, we will try to fix some coordinates within the sort of “small”, further subdivided into three categories: Single, Sophisticated, and Sculptural. Around the concept of small, KiDuk Kim’s movie, the work of Marc Angeli, Carlo Mollino’s drawings, and the crazy lists of Borges will help us to give consistency to the analysis of the small dimension, accomplished microcosmos within which complex events perform.

照片提供：ChongKil Kim

电影《春、夏、秋、冬……和春》（2000年由金基德导演）的背景建筑，位于韩国JuWang山脚下的JuSanji池塘中央

## 一个中心，一份列表，3S：关于“小”的故事

### 作为重心的寺庙

金基德在他的电影中描述了这样一个和尚：他在满是灰尘的林间小道和湿滑的冰层上行走，费力地爬向山顶，身上绑了一个小雕像，还带了一块很重的石头，石头象征着原罪和身体上的忏悔，这样的忏悔也许能使这个和尚赎罪。

在艰难旅程的尽头，主人公惊奇地看到了下面的山谷和围绕在绿色森林里的小湖。小湖中央宁静地坐落着一座建筑，它像一块磁铁一样将整个景观聚拢在一起：这正是主人公出发的那座寺庙，一座漂浮在结冰的平静湖面上的小宝塔。

镜头突然拉近似乎一下子加强了这个空间的特点：这座小型建筑就是天地万物的中心。这是导演金基德拍摄的电影《春、夏、秋、冬……和春》（于2003年）中所描述的景象。在103分钟的电影中，所有的故事都是围绕这个寺庙发生的：尽管有着壮丽的景观、威严的山脉、神秘的水域以及故事的戏剧性力量，人们还是会感受到这个小型的、如漂浮一般的建筑的存在，就如同在电影中压缩的能量得到释放一样。该建筑的大小变成了一个多余元素，对于叙述目的起不到任何作用：在封闭的围墙内，似乎在这些束缚之外的任何情况下都可以找到宁静。从这个意义上说，寺庙的精神层面被放大了，它试图去改变周围的环境。当人们发现这一点时，就如同和尚在山谷之巅俯视全景一样。

从看到的事物中脱离出来，这段能够看清真实事物的相当长的距离将我们带回到和谐的理念中，这种理念在电影中通过事件循环往复的过程体现出来，同时也借助对立面之间连续不断的对话得以展现——罪与赎罪、悲伤和喜悦、焦虑和内心的宁静。因此，我们所需要的，只是当世界似乎变得混沌的时候，可以让我们回归到自然界的某个点。

### 回归到颜色

看韩国导演金基德的电影，关键是要目不转睛地盯着一个点，在电影中这个点就是一座小型宗教建筑。同样的事例也出现在比利时艺术家马克·安吉利的绘画系列作品中，其作品具有一种可视化的抽象概念。马克·安吉利在画廊的白墙上展出了单色系的小型作品，他在木料表面用画笔着色，使其清晰流畅。但它们不是绘画作品，实际上是三维平行六面体，这些六面体的深度揭示了外表面颜色产生的紧凑效果。

这些小型作品聚拢在一起，固定在墙上，小规模的侧面设计能紧紧抓住观赏者的目光。就如同在金基德的电影中一样，浓缩到某一点

## A Center, a List, 3 S: the Tale of the Small

### A temple as a center of gravity

The monk, interpreted by KiDuk Kim, climbs up laboriously to the top of the mountain, between bristling paths of dusty earth and layers of slippery ice. He brings a small statue and, tied to the body, a heavy stone, a symbol of an ancient guilt and of a physical repentance that may perhaps expiate the sins of the monk.

At the end of this difficult journey, almost surprised, the glance of the protagonist settles over the valley below and over the small lake closed in the green forest. At the center, like a magnet around which focuses the entire landscape, a building lies placidly: it is the temple from which comes the monk, a tiny pagoda, floating docilely on the still waters of the lake, now frozen.

A zoom suddenly seems to emphasize this spatial character: the small building is the center of the moderate universe, told in「Spring, Summer, Fall, Winter... and Spring」, directed by KiDuk Kim and released in 2003. Around this temple/house, the cycle of events, that take place in the 103-minute duration of the movie, has been bound: despite the grandeur of the landscape, the majesty of the mountains, the mystery of water, the dramatic power of the events, the small floating building manages to channel in its presence, as compressing, the energies released during the movie. The size becomes a superfluous element, useless for the purpose of narration: within the limits enclosed by the walls, it seems to find peace in every circumstance that occurs outside of these boundaries. In this sense, the spiritual presence of the temple is magnified, managing to change the surrounding territory, as one discovers, like the monk, watching the panorama from the summit above the valley.

The detachment from the observed object, the considerable distance from where it is possible to embrace all the reality, brings us back to the idea of unity, which in the movie is exemplified by a cyclic return of events, by a continuous dialogue between the opposition, between sin and atonement, sorrow and joy, anxiety and inner peace. Very little is needed, therefore: a point, a physical place to which return when the chaos of the world seems to prevail.

### And a return to the color

A key to reading the Korean director's movie is the gaze focused on a point, physically translated in a small religious building.

This happens, again on a visual level but abstract, in the series of paintings carried out by the Belgian artist Marc Angeli, who exhibits on the white walls of the galleries that house his work a series of small paintings, monochromatic. They are pieces of wood with a surface colored with full brushes, making it smooth and clear.

But they are not paintings: they actually are three-dimensional parallelepiped that reveal, with the depth, the straining of the color on the outer face.

马克·安吉利的蓝色体块，2010（左）和白色体块2005（右）
Bleu, 2010(left) and Bloc Blanc, 2005(right) by Marc Angeli

是必不可少的，在这样的情况下，点增加到一定的数量，在白色画廊的背景下就形成了色彩斑斓的星群效果。

因此，尺寸要求这种集中性，而且物体的数量为其增加新的诠释内容：列表（list）。点变得越来越多，使墙壁上出现几何序列，溶解边界并增加深度感。

**建筑分类**

这份列表成为一种力量的汇集体，该汇集体仅凭一个中心点是不能实现的：安吉利的画作通过重复运用成功地赋予作品一种有形的生命气息，其画作带有适度的但很重要的变化：色彩、深度及侧面的尺寸。因此，这个列表形成了一个体系，该体系成为了一种带有有序目的的分类系统。

卡罗·莫里诺和Franco Vadacchino的《建筑、艺术与技术》（1947年）一书中指出，建筑历史的发展伴有微小的存在，这些微小构成了建筑环境中带有注解的列表。这些微小的存在是用铅笔简单勾画而形成的模糊的轮廓，尽管规模很小，但却表现了它们在建筑历史中存在的重要性。

带着些许惊奇，我们可以看出，不太著名的建筑虽描绘简洁但特性鲜明：人们不禁会想，是否是当地环境衬托了建筑形象，而不是建筑形象衬托了当地环境。这如同读一本书，像中世纪的条约，或后来出现的启蒙条约，即一个人可以想象通过哲学语言将已知的世界记录在册（其中每一个单词都是由一组字母或音节组成的，暗指将已知的世界归类）。

默里诺和Vadacchino的表达形式是建筑，在由几笔勾勒而成的封闭空间中，迫使我们在那个所描述的微小世界中集中注意力，相应地，也能够使我们在轮廓的界限范围内想象建筑的空间。

**因此谈到博尔赫斯：关于分类**

豪尔赫·路易斯·博尔赫斯向我们描述到：约翰·威尔金斯在17世纪中叶发明了一种哲学语言或分析语言，该语言使我们看到《其他调查》（1952年面世）作品集中《约翰·威尔金斯的分析语言》的一小段文章中描述的一些惊人的列表。

这些含糊不清的话、赘词和不足的情况使我们想起弗朗茨·库恩博士在一本中国百科全书《天朝拾遗录》（Celestial Empire of Benevolent Knowledge）中的提法。该文后面页码的记载中将动物划分为：（a）属于帝王的，（b）用防腐药物保存（尸体）的，（c）驯化的，（d）乳猪，（e）女妖，（f）寓言中的，（g）流浪狗，（h）包含在当前的分类中的，（i）狂热的，（j）不计其数的，（k）用精制的驼毛刷绘制的，

These small volumes take strength in being together, fixed to the wall; the small size of the sides compels the viewer to an effort of attention. As in the movie of KiDuk Kim, a concentration that converges to a point is required; in this case, the point increases in number, becoming minute colored constellations on the white background of the gallery.

The size, therefore, requires the focus and the number of objects adds a new interpretation: the list. The points become more numerous, giving a geometric order on the wall, dissolving boundaries, and adding depth.

Architectural taxonomy

The list becomes the construction of a force that the focus on one center can not reach: the paintings of Angeli succeed in giving life to an almost physical aura with a sort of exercise of repetition, with modest but significant variations: the color, the depth, the size of the sides. It then builds a system that becomes taxonomic with an ordered purpose.

In the book by Carlo Mollino and Franco Vadacchino in 1947, *Architecture, Art and Technique*, the flow of a history of architecture is accompanied by small presences, making up an annotated list of buildings to accompany the text. These are slight figures, dotted with a few lines in pencil, which can communicate a great thoroughness, despite the tiny dimensions.

With stupefaction, we can recognize architectures, more or less famous, portrayed with terseness, but with a clear identity; one wonders whether it is the text to accompany the images, rather than vice versa. It seems to read, between the pages of the book, like a medieval treaty, or by sliding forward over the years, an Enlightenment treaty, when one could imagine being able to catalog all the known universe through philosophical language (in which each word is composed of a set of letters or syllables that refer to a division of the known world into categories).

The language of Mollino and Vadacchino is architecture, in the complexity of an enclosed space within a few drawn strokes, which forces us to bring the sight within that depicted little world, which, in turn, allows us to imagine the void defined within the limits of the figure.

Thus spoke Borges: on taxonomy

Describing the figure of John Wilkins, who, in the mid-1600s, invented a philosophical language, or analytic language, Jorge Luis Borges involves us in some amazing lists, described in the short text *The Analytical Language of John Wilkins*, which is in the collection *Other Inquisitions*, which appeared in 1952.

*These ambiguities, redundancies and deficiencies remind us of those which doctor Franz Kuhn attributes to a certain Chinese encyclopaedia entitled Celestial Empire of Benevolent Knowledge. In its remote pages it is written that the animals are divided into: (a) belonging to the emperor, (b) embalmed, (c) tame, (d) sucking pigs, (e) sirens, (f) fabu-*

由雷姆·库哈斯和布鲁斯·茅尔所著的《小，中，大，超大》，Monacelli出版社，1995年
*S, M, L, XL* by Rem Koolhaas and Bruce Mau, Monacelli Press, 1995

(l)其他的，(m)刚刚打破水罐的，(n)来自很远的地方、看起来像苍蝇的。

在这个令人惊讶的列表中出现了一种可控的"疯狂"，这种"疯狂"迫使我们重新思考当我们在面对一系列事物和对象时所要进行的日常分类和编目的操作方式。我们在观察莫里诺的小画像、安吉利的白墙画廊上零散着的绘画作品和在韩国山谷里消失的寺庙时也会运用同样的思维方式。

**因此，库哈斯区分了大小**

《小，中，大，超大》这本书根据建筑规模组织建筑材料，但这些建筑材料之间并不是紧密联系在一起的。书中有很多论作都穿插到了项目中，但这些论作并不与项目完全吻合，而是作为独立的片段而存在的，所以书中很多的矛盾之处是无法避免的。不管怎么说，这本书是值得一读的。

由雷姆·库哈斯和布鲁斯·茅尔所著的《小，中，大，超大》一书于1995年出版发行，该书通过大量文字和OMA工作室设计的项目反映了这一主题。

突然间，对建筑主体的研究不再依据形态或类型学的问题，也不基于功能和结构的要求，抑或是建筑材料，而是按照实物大小将建筑分类：一种既陈旧又创新的分类。从那以后，例如"大"本身就变得有价值了，成为一个空间的特征元素，而且"大"是独立的，因此空间将容纳某一功能而不是另外一种功能。

由库哈斯推出的分类假定将建筑学变为一种测量列表，该列表使其重回到一种实物测量模式或许是更人性的模式，因为即便不是专家也能轻易地推论出：视线、触觉、平衡感实际上可以成为立体分析的对象，可以直接由参观者比较出它们的不同。

库哈斯将我们带回到由金基德拍摄的小型建筑所体现出来的价值观中："小"与"中""大""超大"有着同样的连贯性，并且因为这个原因，"小"极大地影响着周围的环境。

**无止境地分割**

然而，暂且不管库哈斯的创新理论，"小"并不是建筑的最终测量标准，因为在微观物理学中，它还未被实验证实是最终的分割实体：我们从原子发展到电子、质子、中子，然后是夸克，而且我们也在研究将其继续细分。此处所列的项目提到的"小"，作为一个类别仍是可持续研究的，它本身可以有更多的分支。

这一系列的发展阶段把我们带到最后这一列表：金基德已经向我们展示了小规模的价值以及通过极小的变化改变一个场所的空间的能

*lous, (g)stray dogs, (h)included in the present classification, (i)frenzied, (j)innumerable, (k)drawn with a very fine camelhair brush, (l)et cetera, (m)having just broken the water pitcher, (n)that from a long way off look like flies.*

A sort of controlled madness appears in this formidable list, a madness that compels us to rethink the daily operations of sorting and cataloging that we do whenever we encounter a series of objects, of beings. It is the same afterthought that we implement in observing the small portraits of Mollino, the paintings scattered throughout the white walls of Angeli, the lost temple in the Korean valley.

**So Koolhaas: on size**

*S, M, L, XL organize architectural material according to size; there is no connective tissue. Writings are embedded between projects not as cement but as autonomous episodes. Contradictions are not avoided. The book can be read in any way.*

In 1995, the book *S, M, L, XL*, written by Rem Koolhaas and Bruce Mau, was published; a substantial reflection on texts and projects by studio OMA.

Suddenly, the research on the body of the architecture is no longer based on morphological or typological issues, is not based on functional and structural requirements, nor on the materials. The architecture is categorized on the basis of the physical dimension: a classification as banal as revolutionary. Since then, Bigness, for example, becomes a value in itself, a characteristic element of a space, and it is independent that the space will accept a certain function rather than another.

The taxonomic hypotheses introduced by Koolhaas turn architecture into a list of measures, reviving its existence to a physical dimension, maybe more human, because it is easily inferable also from non-experts: the sight, the touch, the sense of balance become, in fact, subjects of spatial analysis, an immediate comparison made directly with the body of the visitor.

Koolhaas brings us back to the values expressed by the tiny building filmed by KiDuk Kim: the "small" has the same consistency as the "medium", the "large" and the "extra-large" and for this reason can strongly influence the surrounding environment.

**Endlessly splitting**

Yet, despite the revolutionary thesis of Koolhaas, "small" is not the ultimate measure of architecture, as in microscopic physics, where it still has not been verified experimentally as an entity of last division: from the atom, we moved to the electrons, protons and neutrons, then quarks, and we search again for a still further division. The projects presented here speak of "small" as a category still investigable, that allows more subdivisions within itself.

A series of steps took us to this last list: KiDuk Kim has shown us the value of small size, the ability to alter the space of a place through minimal changes; Angeli worked on the multitude of

由铜和玻璃构成的悬挂结构—茶室，马里兰，美国
A hanging bronze and glass object, Tea House, Maryland, USA

由铜和玻璃构成的悬挂结构—茶室，马里兰，美国
inner court of Gallery Aso, Daegu, Korea

力；安吉利致力于大量实体的研究，这些实体通过增加每一个微小部分的变化以及它们的相互作用而产生，最后使之成为一个栩栩如生的列表；莫里诺和Vadacchino将我们带回到建筑学当中，创建了一个汇集复杂性的小缩图列表，该列表改变了分类系统里的列表。博格斯再次打开了人们的视野，促使我们回忆起列表所产生的清晰的疯狂；而且，库哈斯最终做出了总结，在一个合理排列的目录中使"小"的价值得以回归。

现在"小"可以自我划分，延伸种类并创建各类之间的关系。以下三个S可以界定所提及到的项目：单一型（Single）、复杂型（Sophisticated）、雕塑型（Sculptural）。

**定义：探索"小"**

单一型

在这个定义范围里聚集的项目，通常都由独特的空间和单人间组成，其清晰可辨的功能界定了整个建筑结构。

属于这个类别的建筑有：A1 Architects的"帽子茶室"、David Jameson Architect的"茶室"、Ravnikar Potokar Architectural Office的"树屋"。这些作品在结构和运动中有一些复杂性——尤其是Jameson的"茶室"，其显示了一个巨大结构上的成就，与空间的规模相比，"茶室"努力将房间与周围的环境隔离开来，这个尝试暗示茶道礼节成为一个与世俗脱离的标志。通过架高大型金属入口，茶道艺术完美地体现了泰然自若与宁静安详。与"茶室"不同，"帽子茶室"借助天空的力量，屋顶上的大型帽子设计及其与周围环境的融合是灵感的来源。墙壁是可以移动的，内外结构混合在一起：小房间可以瞬时变成庞大的房间，在一个有趣的盖板下可以看到一片广阔的风景——这与"树屋"看到的风景一样，单从视觉角度看，它并没有因其孤立性而悬浮在地面上，而是更好地与地面结合在一起。

**复杂型**

复杂与单一是相对的，在这种情况下，"小"几乎失去了意义，因此，高度体现了以下所列项目中封闭空间的复杂性：5468796 Architecture的"立方体"、BNKR Arquitectura的"日落小礼拜堂"、Andrea Tognon的"住宅O"、Shinichi Ogawa & Associates / urbanist architect的"立方体住宅"、Architecture H的"Aso展廊"和el348Arquitectura的"圣安妮小礼拜堂"。

"日落小礼拜堂""Aso展廊"和"圣安妮小礼拜堂"似乎是从之前描述过的单独空间中演变发展而来的：在这些冥想的空间里，功能都是独特的，无论是博物馆还是宗教场所，所有的组成部分趋向于使这

objects generated by adding the change from every little piece with the interaction between them, making a live list; Mollino and Vadacchino have brought us back to architecture, creating a list of small microcosms that, dense of their complexity, have transformed a list in a taxonomic catalog; Borges opened the horizons again, forcing us to reflect on the lucid madness generated by lists; and Koolhaas finally has closed the circle, restoring the value of the "small" within a catalog raisonne.

Now the "small" can divide itself, expand its own description, build relationships. Three S's could limit the projects presented: Single, Sophisticated, Sculptural.

**Definitions: exploring the "Small"**

Single

The projects collected within this definition generally are constituted by unique spaces, single rooms that delimit the entire architecture within a well-defined function.

Falling into this category are the Hat Teahouse by A1 Architects, the Tea House by David Jameson Architect, and the Tree House by Ravnikar Potokar Architectural Office. They are works that find complexity in the structure and in the movement: in particular, the Tea House by Jameson tells of a huge structural effort, compared to the scale of space, a necessary effort to isolate the room from its surroundings, suggesting a ceremony that becomes almost unworldly. Raised by large metal portals, the tea ceremony is, ideally, undisturbed and unperturbed. Unlike Tea House, Hat Teahouse, which invokes in the sky, thanks to the big hat holed on the roof, a source of inspiration, as well as in the surroundings. The walls become mobile and the exterior is mixed with the interior: the small room suddenly can become gigantic, welcoming under the funny cover a broad landscape – the same landscape that looks for the Tree House, not suspended above the ground for isolation but for a greater integration, if only visually.

**Sophisticated**

Sophisticated is an opposition to the single: in this case "small" almost becomes meaningless, so high is the complexity of the space enclosed within the projects listed below: The Cube by 5468796 Architecture, the Sunset Chapel by BNKR Arquitectura, the Residence O by Andrea Tognon, Cube House by Shinichi Ogawa&Associates/urbanist architect, the Gallery Aso by Architecture H and the Saint Anne's Chapel by e|348 Arquitectura.

The Sunset Chapel, the Gallery Aso and the Saint Anne's Chapel seem evolutions of the individual spaces described before: places of contemplation where the function is unique and all their parts converge in the aim to give more prominence to this function, whether a museum or a religious space. Yet here the singularity evolves, the spaces multiply, some paths are born. Contemplation is less absolute; it is "dirty" by a richer collection of views, of spaces.

十字门，Dukovany，捷克共和国
The Cross-gate, Dukovany, Czech Republic

照片提供：Ivo Pavlik

木柴堆，温尼伯湖，加拿大
Woodpile, Winnipeg, Canada

照片提供：Biran + Talmon(©Brian Gould)

些功能更加突出。然而，在这里单一性演变发展，空间得以增加，小路应运而生。冥想变得不那么绝对；空间的广阔视角使这里不再纯粹。

这种效果甚至在两个住宅建筑“住宅O”和“立方体住宅”中也能体现出来，然而，包含在普通住宅里的功能在数量上的增加，产生了一系列具体的房间。在这样的情况下，“小”的定义就体现在复杂的综合体中，体现在维护性上，这些都在界定的范围内得以实现，还体现在清晰可辨的外形上，如同“立方体住宅”最重要的平行六面体一样。

“立方体”是一个小型的可移动的珍宝，其中舞台功能用三维金属折叠帐篷建成。这种立方体虽然完美地包围着舞台，却也像城市中心跳动的心脏，是封闭的。影像和光线通过墙壁向外投射，并反射到城市周围，使城市的景色生动起来。

### 雕塑型

这个词突出了建筑和艺术作品之间微妙的界定。它既不是界定建筑的内部空间，也不是一种功能：它是接近环境与人的一种方式。在以下列出的项目中，与景观完全不同的建筑外形成为参观者和建筑之间的第一个联结方式。在这种定义里，我们可以看到Noa Biran和Roy Talmon的“木柴堆”、Petra Gipp Arkitektur的“避难所”和Ivo Pavlik的“十字门”。

“木柴堆”和“十字门”的特点是使用粗糙的天然材料。堆积的木块燃烧后就消失不见了，打开装有它们的笼子或再次火烧干草，留下的是它曾存在于混凝土上的记忆。它们是人类的作品，但似乎是直接来源于主宰它们的自然环境。“避难所”则相反，除与该区域融合外，还以对立的方式寻求一体化，在这个意义上，围绕轻木体量周围的白墙传递了一种景观内充满活力的意识，这种意识又自相矛盾地回归到自然当中。

### “小”是……

和尚在他的童年时期，在巨大的佛像肩膀上凝望着家乡——湖中心如漂浮一般的小寺庙。他在寻找安全感，即一个将他的存在置身于他周围的大自然环境中的中心点。多年后他会在山顶上发现这个点，当一座小型建筑作为一座“建筑”展现自身时，它就能够改变主人公眼中所出现的整个画面。

Diego Terna，设计师，是网络杂志PresS/Tletter“想象建筑”部门的负责人。他在意大利的米兰和西班牙的圣塞巴斯蒂安学习过建筑设计，他赢得过由意大利评论家Luigi Prestinenza Puglisi举办的青年评论大赛奖，从那以后他的作品出现在多个艺术和建筑杂志中（Compasses, Arch'it, Exibart）。他在Boeri工作室积累了工作经验后，于2005年开始一直在伊塔洛罗塔工作室工作。

This happens even in two residences, Residence O and the Cube House, where, however, the functions contained within the generic one of living increase in number, into a series of specific rooms. The definition of "small" in this case is found in the unity of the complex, in the maintenance, within limited dimensions, of shapes clearly discernible, as the supremacist parallelepiped of the Cube House.

The Cube is a small mobile jewel, in which the function of the stage is built with folds of the metal three-dimensional tent. But the perfect cube enclosing the stage also lives closed, like a beating heart in the city. Pictures and lights flow through walls, projecting outward, reverberating in the urban surrounding, giving life to the urban landscape.

### Sculptural

In This word is highlighted the tenuous border between architecture and art work. It is neither the interior space to define a building as architecture nor it is the function: it is an approach to the surrounding and to the people. In the projects listed below, the shape of the building, its being completely different from the landscape, becomes the first relationship between visitors and the building. In this definition, we find the Woodpile by Noa Biran + Roy Talmon, the Refugium by Petra Gipp Arkitektur, the Cross-gate by Ivo Pavlik.

The use of rough natural materials characterizes the Woodpile and the Cross-gate: pieces of wood, stacked, which disappear as long as they are burned, unlocking the cage that contains them or, again, the fire burns the hay, leaving forever the memory of its presence marked on the concrete. They are works of man, but seem to derive directly from the environment that hosts them. The Refugium, on the contrary, searches an integration using opposition, going in addition to the territory; in this sense the white walls, twisted around the net volume of light wood, transmit a sense of vitality in the landscape, returning paradoxically to nature.

### Small is

On the shoulders of a large Buddha, the monk, during his childhood, sights his glance to his home, that little temple floating in the midst of the waters of the lake. He is looking for safety, a point that centers his existence in the huge landscape surrounding him. He will find it years later, in the peak of the mountain, when the small building will show itself as an "architecture", able to change the whole scene that takes place under the eyes of the protagonist. Diego Terna

**Diego Terna**, architect, curator of the section "Imagined Architectures" in the webzine PresS/Tletter. Studied architecture in Milan, Italy, and San Sebastian, Spain. After he won a young critics competition, called by Italian critic Luigi Prestinenza Puglisi, his writings have appeared in several art and architecture magazines (Compasses, Arch'it, Exibart). Has been working for Italo Rota office since 2005, after experiences at Boeri Studio.

# 住宅O

Andrea Tognon

我们要翻新的这座建筑大楼建于20世纪70年代，模仿了威尼托乡村的本土建筑风格。现存的建筑看起来相当地不完整。建筑的平面图是正方形的，但拐角已经不见了，所以该建筑呈现L形。屋顶是一块混凝土厚板，向外突出很不美观，也不成比例。

我们决定补上缺少的拐角，使建筑平面图变得完整。由于重新界定了保温参数，我们重塑了建筑的外形：切掉了以前屋顶上的边缘，并重新设计了屋顶和外墙间的连接点。整个内部布局都被重新设计了，所有的墙体和屋顶都加上了保温层，采暖系统改成了太阳能采暖。

建筑几何形状的重新定义让人们对这里产生了全新的认识，一个简单的住宅原型变成了一个可以俯视无限美景的纪念碑式的建筑。建筑本身没有任何临时建筑的意味。这项研究不是以最简的形式主义命名，而是本着安静、简单的精神，使生命呈现出宁静、审慎的一面。

## Residence O

The building we were called to refurbish was built in the 1970's as an imitation of vernacular architecture of the Veneto countryside area. The existing building looked pretty incomplete. The floor plan was a square but the corner was missing. So the building was in an L shape. The roof was a concrete slab jutting out in a very inelegant and bad proportional way.

We decided to add the corner that was missing to complete the square floor plan. Because of the total redefinition of the insulation parameter, we reshaped the profile of the building by cutting the old roof edge and redesigned the junction between roof and exterior walls. The entire interior layout was redesigned, all the walls and roof insulated, the heating system switched to solar energy.

The redefinition of the geometry of the building brings a new radical perception of the place, where the archetypal simplicity of the house becomes a monumental object overlooking a timeless landscape. The building in itself erases any temporary connotation. The research is not by the name of a minimal formalism, but in the spirit of a quiet simplicity, where the presence of life tries to be quiet and discreet. Andrea Tognon

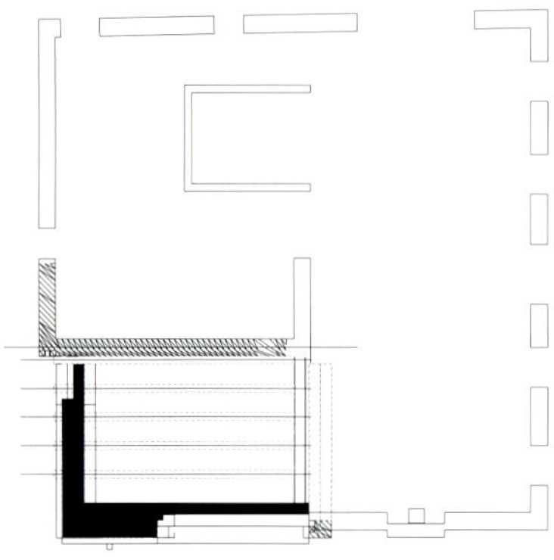

二层——改造前
second floor _ before

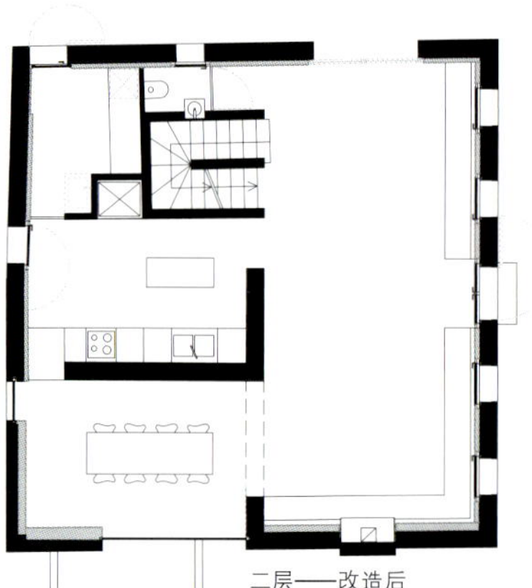

二层——改造后
second floor _ after

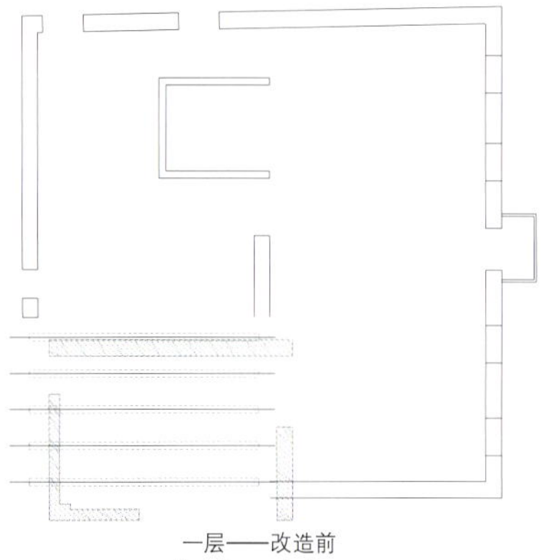

一层——改造前
first floor _ before

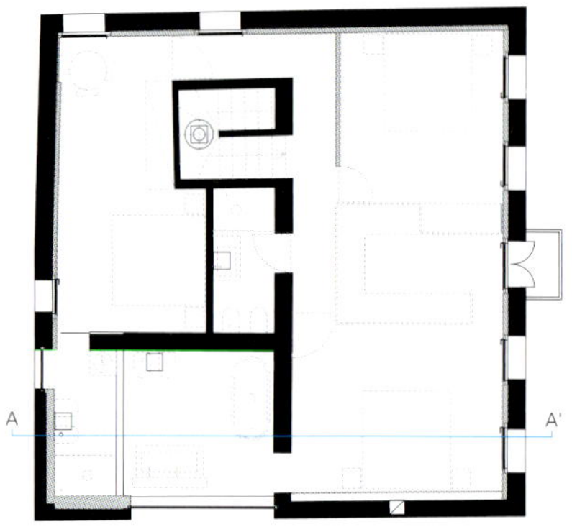


一层——改造后
first floor _ after

屋顶 roof 0 1 3m

项目名称：Residence O
建筑师：Andrea Tognon
合作者：Roberta Cazzaniga
结构工程师：Enrico Savorani
电力工程师：Giancarlo Zaghetto
承包人：Faulisi
甲方：Mr. O
地点：Teolo, Italy
基地面积：3,900m² 建筑面积：110m²
结构：Faulisi
外部装修：Stone
内部装修：Concrete
材料：wood, concrete, stone
设计时间：2007.8
施工时间：2009~2010
竣工时间：2010

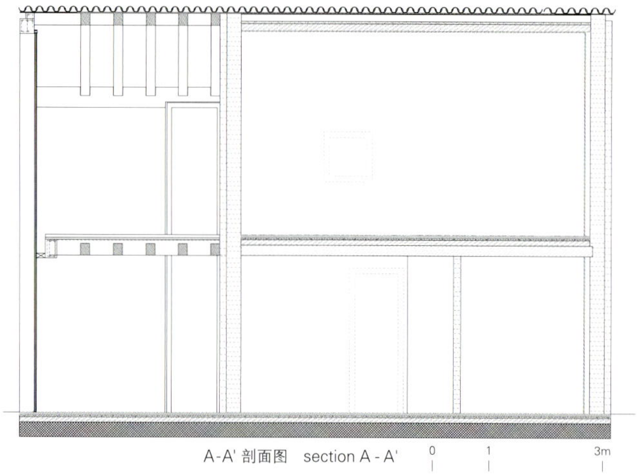


A-A' 剖面图　section A - A'

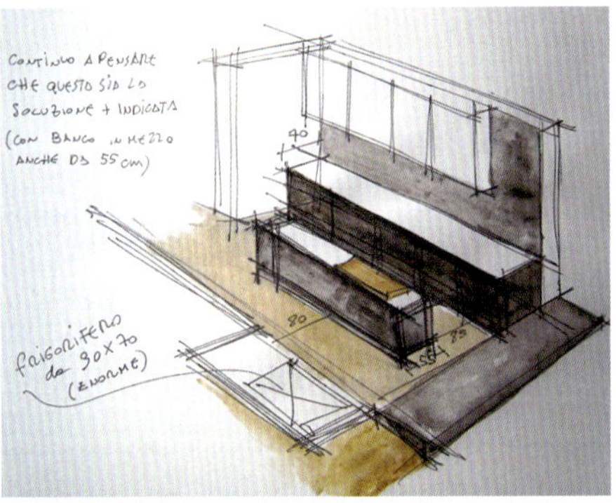
CONTINUO A PENSARE
CHE QUESTO SIA LA
SOLUZIONE + INDICATA
(CON BANCO IN MEZZO
ANCHE DA 55 cm)
40
80
35
frigorifero
da 90x70
(ENORME)

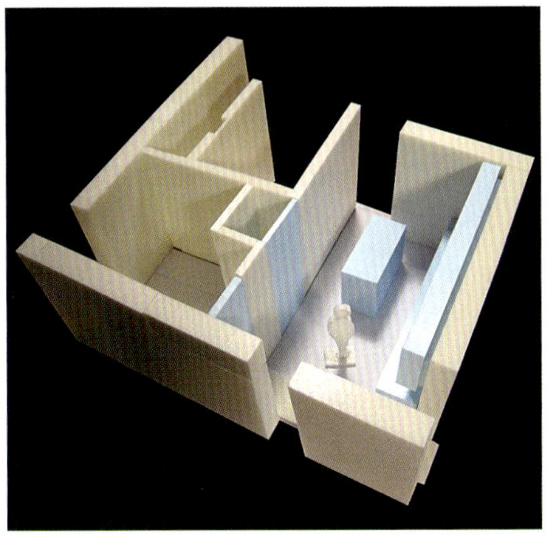

# 圣安妮小礼拜堂

e|348 Arquitectura

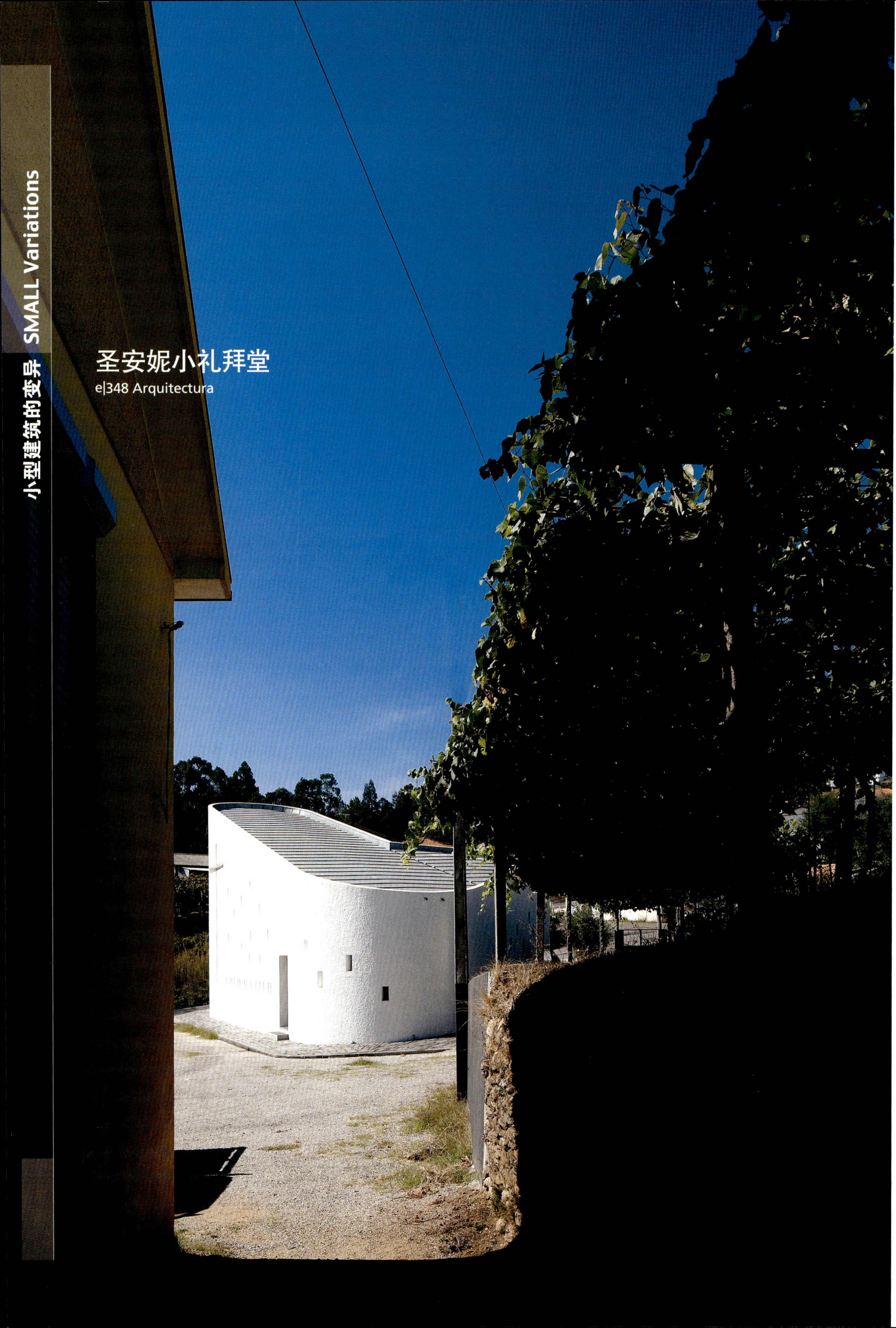

2009年初，我们接受委托建造一座纪念圣安妮的小礼拜堂，项目基地位于圣玛丽亚费拉Sousanil小镇的一小片三角形土地上。

我们走访了这块土地，发现小礼拜堂的位置处在五条街道的交汇点上，末端拐角地势较高。我们还注意到当地人每年7月26日都会举行庆祝圣安妮节的欢庆活动，而这种地形条件则为庆祝活动提供了一个完美的“舞台”。这种圆形的露天剧场外形使人人都有极佳的视角，这更加鼓舞人们参与到活动中来。

四周的环境是该项目的关键所在。这座建筑的修建使得这种自发的圣地所带来的熏陶与启迪得以实现，改善了庆祝条件，同时也为纪念圣安妮创建了一座卓越的建筑。

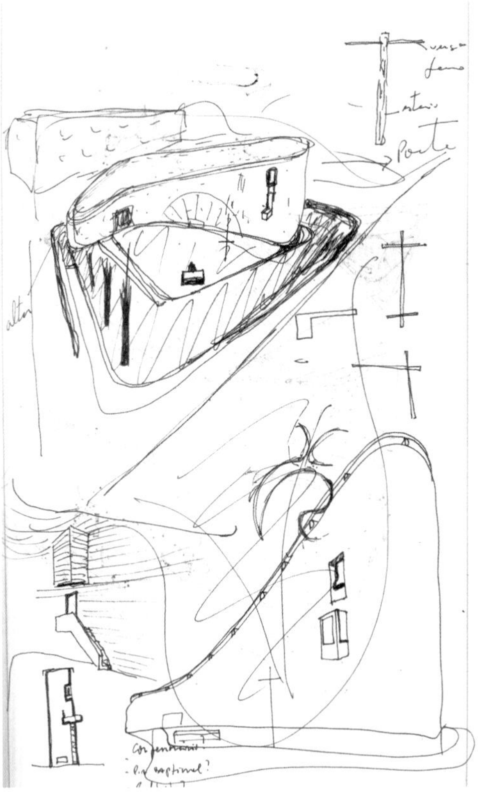

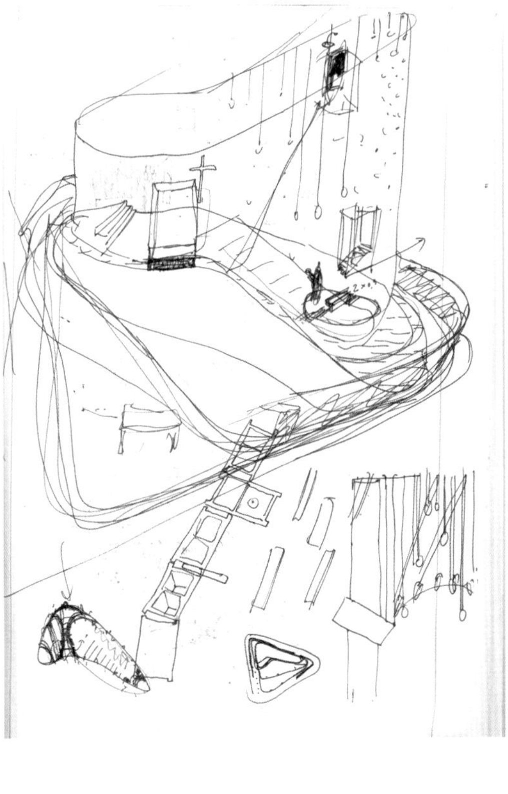

在我们的脑海中有两座主要的建筑参照物：一座是圣保罗教堂，建于12世纪的一座非凡的小型古罗马建筑，位于塔沃拉河旁的一座花岗岩斜坡上；另一座是朗香教堂，它是近年建筑史上的一个建筑参照物，我们将其解读为自然景观中的标志建筑，它采用了雕塑式的外形、粗糙的材料和饰面，还严格限制了光线的进入。

这样的设计也激发了我们对建筑与外部空间之间联系的设计灵感。人们常在建筑外举行大规模的庆祝活动，这实际上就是墙体有机形态形成的主要原因，这样的墙体能成为自然的音响设备，有助于声音的传播。

圣安妮小礼拜堂的三角形场地恰好适合创建一座L形的建筑，这样能够很好地组织外部的空间，并妥善地处理内部的建筑功能。

该项目只有一层，只是祭坛的高度与教堂顶部的高度不同。它为室外大型庆祝活动创建了空间，人们可看到周围的大量美景。如果把它当作一个礼拜的场地或一座卓越的建筑，它就可以在周围的环境中成为一座强有力的视觉参照物。

正门面向西方，进入“小礼拜堂门厅”，可以看到圣安妮的雕像，这里是为人们提供首次礼拜和祷告的地方。再往里走，就进入到一个可容纳30人的教堂中殿，后面设有祭坛。

引导此设计的美学原理旨在增加人们的宗教体验，这是通过宗教与建筑灵感的最大源泉——光来实现的。像圣保罗教堂一样，小礼拜堂的空间是封闭的，巨大的墙体上设置了狭窄的洞口。这种控制光线的做法不仅参考了只有七扇窗的圣保罗教堂，同时也参考了朗香教堂的内部设计，其门窗洞口的布局创造了一个适宜的氛围，让游客可以更好地冥想与思考。

小礼拜堂的室外空间形成了自发的露天剧场，使观众可以进行室外庆祝活动。参照朗香教堂后，我们在钟下和观众席上设置了楼厅，使其成为建筑中唯一向外延伸的部分，从外面看几乎像一件雕塑品。

圣安妮小礼拜堂带着一些人的强烈愿望得以建成，在这里人们可以聚集在一起做礼拜，它连结了历史与现在，行使着其宗教目的，最重要的是促进了城市的进步和社会的凝聚性。

## Saint Anne's Chapel

In the beginning of the year 2009, we were commissioned the project of a Chapel honoring Saint Anne, in a small, triangular piece of land, in Sousanil, Santa Maria da Feira.

Visiting the site, we noticed that the given location of the Chapel was in the intersection of five streets/roads, and at the bottom corner of an ascending topography. We also noticed that this topographic condition was the perfect "stage" for celebrating Saint Anne's festivities, occurring on July 26th every year, as locals already do. The amphitheater shape of the site allows everyone to have optimal visibility conditions, promoting their participation in the ceremonies.

This circumstance was the leading key for the entire project. The building to develop would materialize the edification of the bottom edge of this spontaneous sanctuary, improving the conditions for celebration along with the creation of a singular building capable of housing Saint Anne's devotion.

In mind, there are two major built references: the São Pedro das Águias Church is a small singular Romanic building of the XII century, built next to the Távora river, in a granitic slope. Notre Damme du Haut, is a reference in the recent architectural history, and we read it as an iconic building that spots the landscape, making use of sculptural forms, and rough materials and finishes, with well measured limitation to light entrance.

It is also inspiring the relationship with the exterior space. Mass celebrations took place outside the building, which is in fact, the

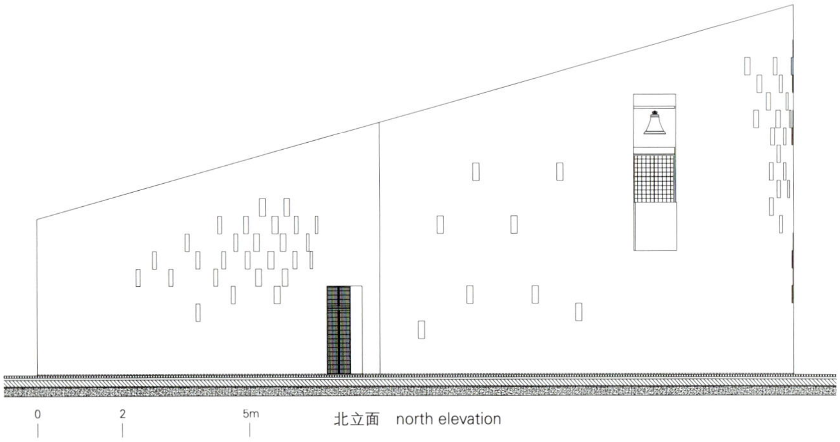


北立面　north elevation

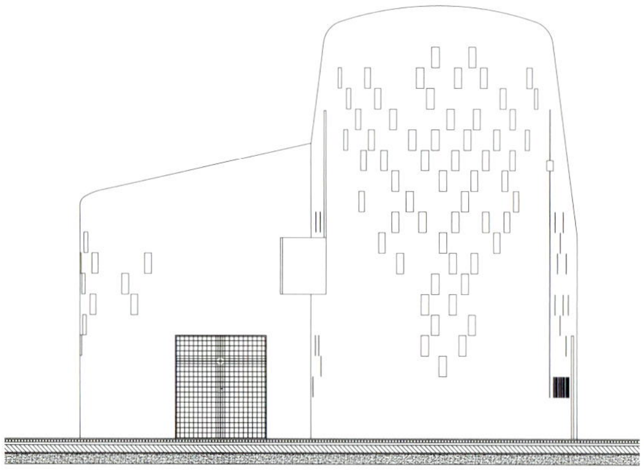

西立面　west elevation

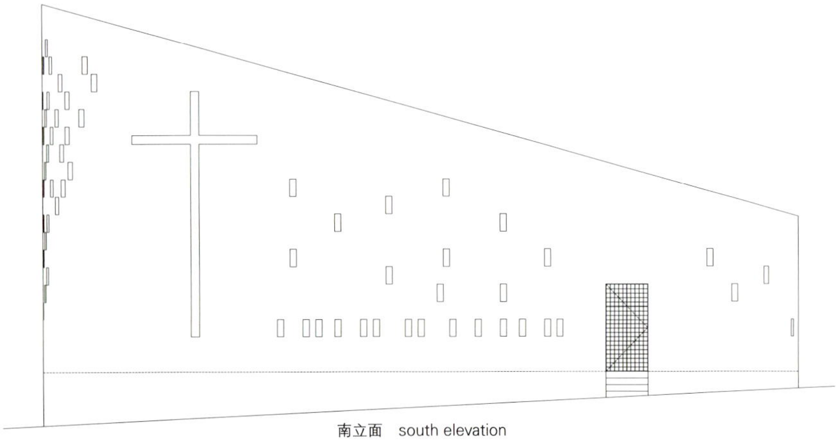

南立面　south elevation

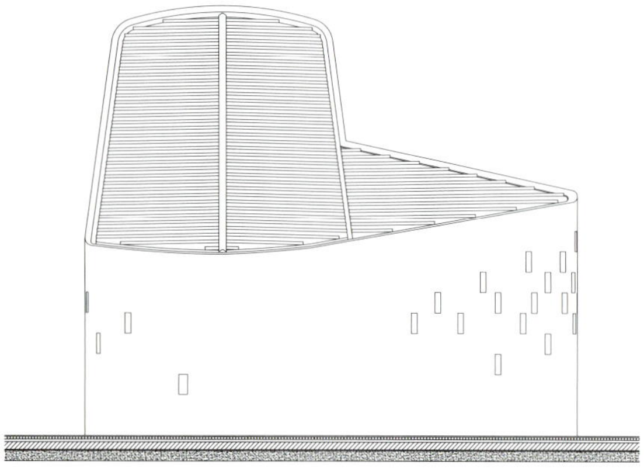

东立面　east elevation

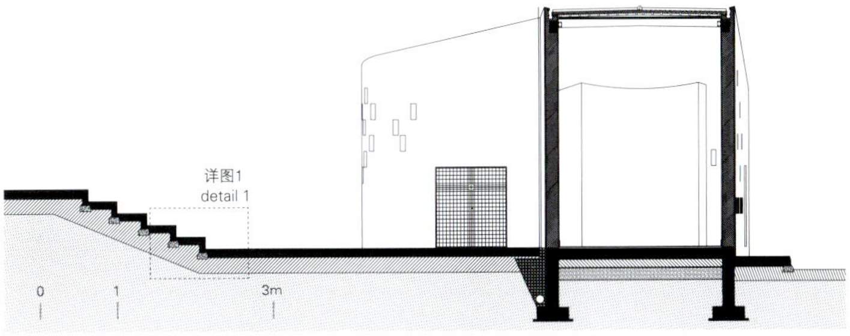

A–A' 剖面图
section A-A'

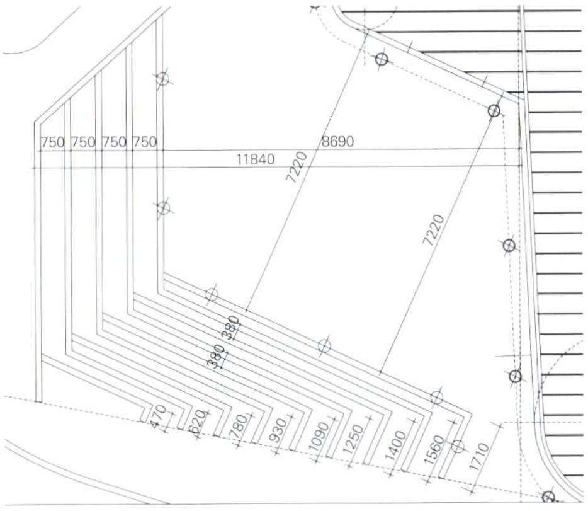


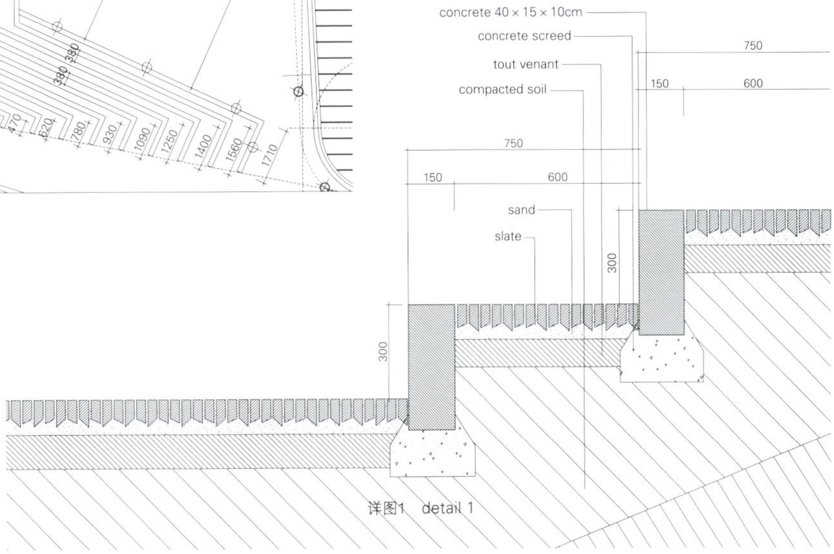

详图1 detail 1

main reason for the organic shape of its walls – acoustics, sound propagation.
The triangular ground given to Saint Anne's Chapel, suggested the creation of an L shaped building, allowing to organize the exterior space, and to dispose properly the inside building program.
It is developed in one level, varying only its height towards the altar and the head of the chapel. It creates a Scenario to the outdoor mass celebration, and visually scores landscape, assuming, as befits a place of worship, as an exceptional building, creating a strong visual reference towards its surroundings.
The main entrance is oriented towards west. Entering, an "ante-chapel" where the image of Saint Anne is located, creating a space for worship and devotion on the first moment. After, we reach the nave, with a capacity of about 30 people, with the altar in the background.
The aesthetic principles leading the design look for the magnification of the religious experience by working the largest source of religious and architectural inspiration, the light. So, as São Pedro das Águias, the space is all closed, massive ascending walls punctuated with thin fenestrations. This light control not only refers to São Pedro das Águias, where there are only seven windows, but also to Ronchamp's interiors, where the distribution of fenestration creates the correct ambiance to elevate visitors to a meditation and contemplation level.
The exterior space materializes the spontaneous amphitheater as to accommodate audience in exterior celebrations. Looking at Notre Damme du Haut, we created a balcony under the bell and over the audience, becoming the only extension of the building towards its exterior– an almost sculptural moment.
Saint Anne's Chapel is the result of some people's strong will to build a place for devotion that is able to gather together people around a place, connecting history and present, serving its religious purpose, and, most of all, serving urban enhancement and social cohesion. e|348 Arquitectura

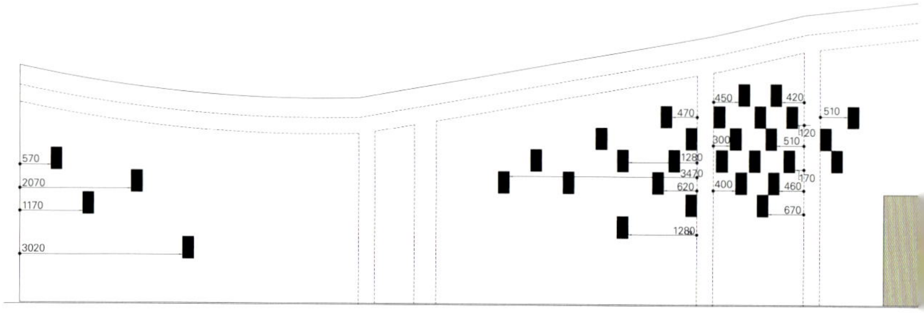

1. kinky finish
2. plaster
3. masonry (concrete block 30 × 20)
4. glass 6mm
5. bitumen / silica

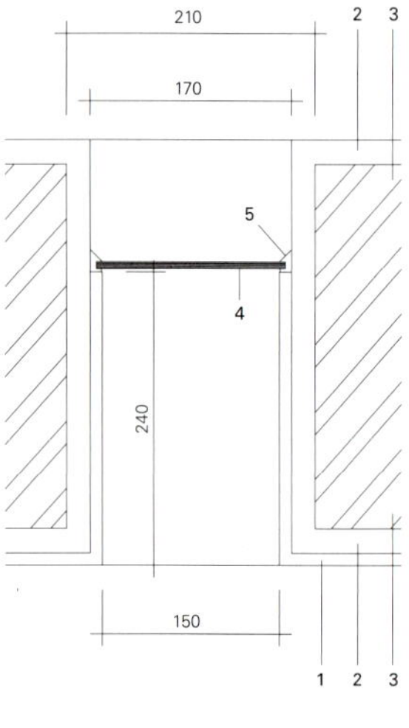

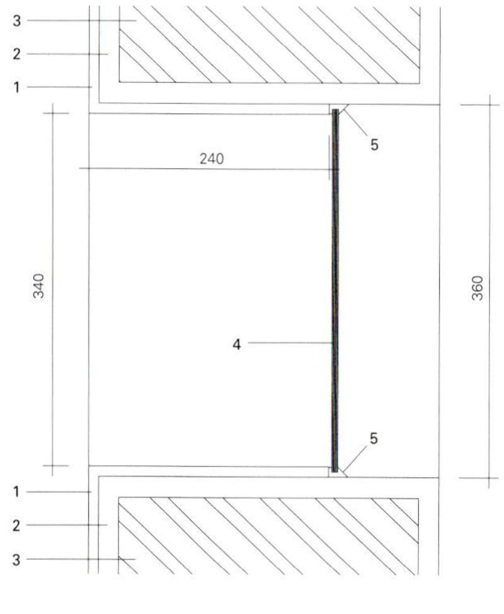


窗户详图
window detail

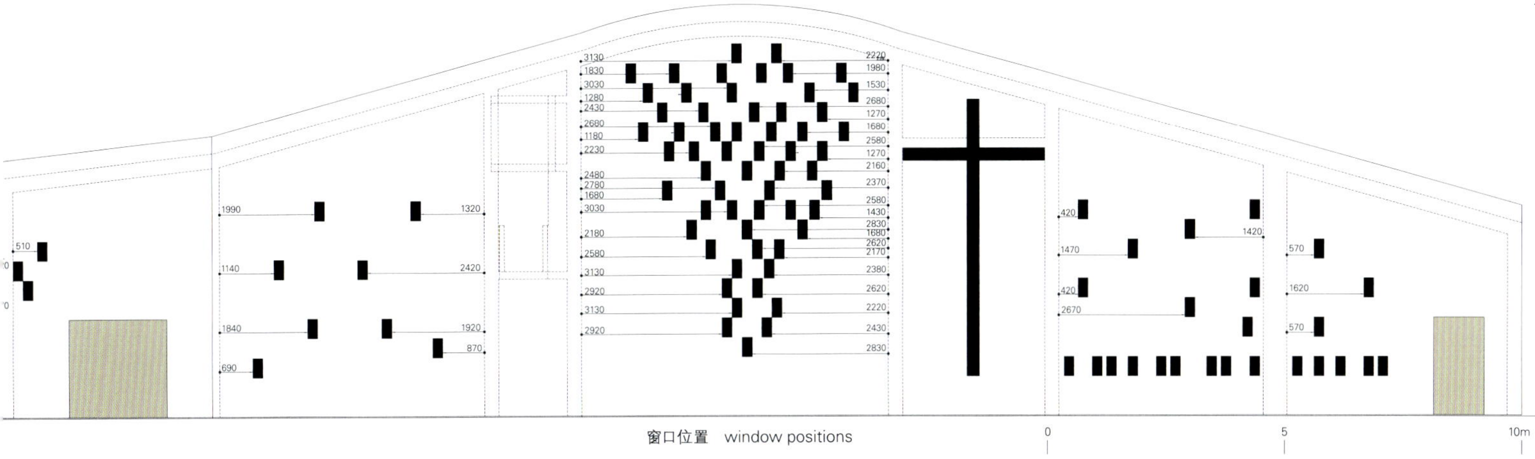


窗口位置　window positions

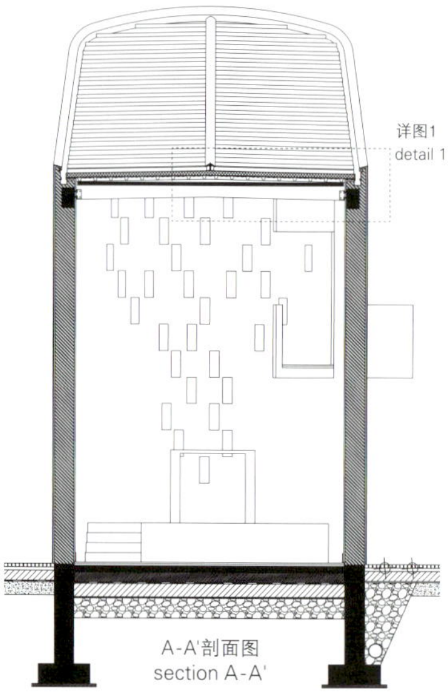


A-A'剖面图
section A-A'

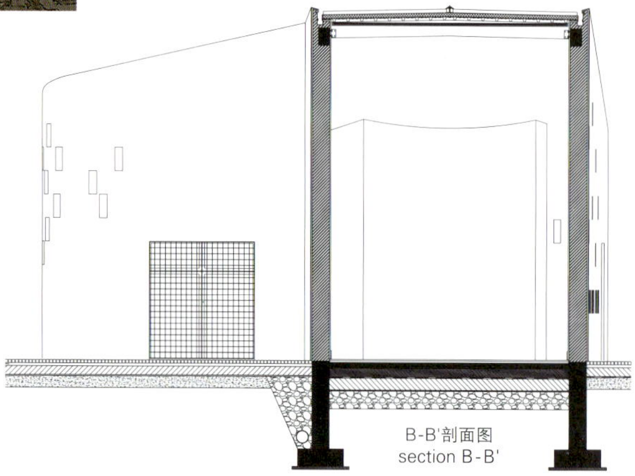

B-B'剖面图
section B-B'

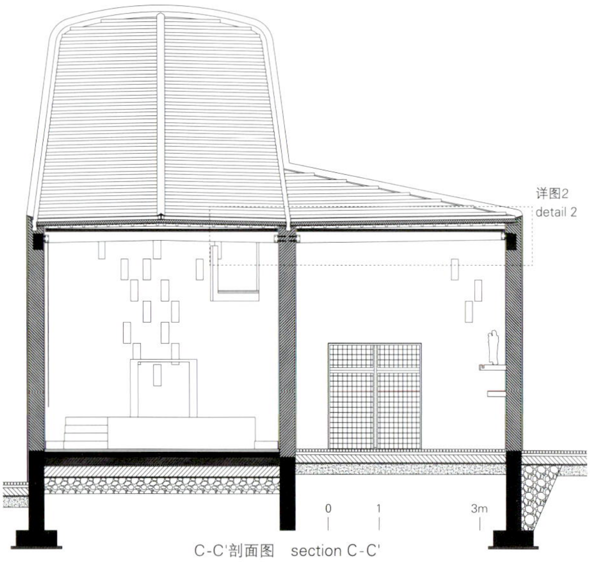


C-C'剖面图 section C-C'

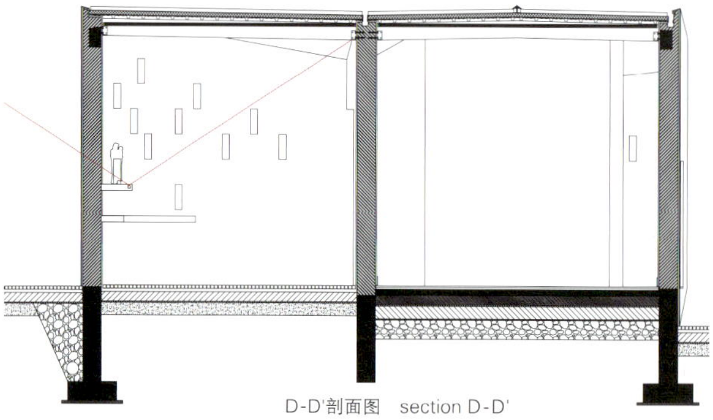

D-D'剖面图 section D-D'

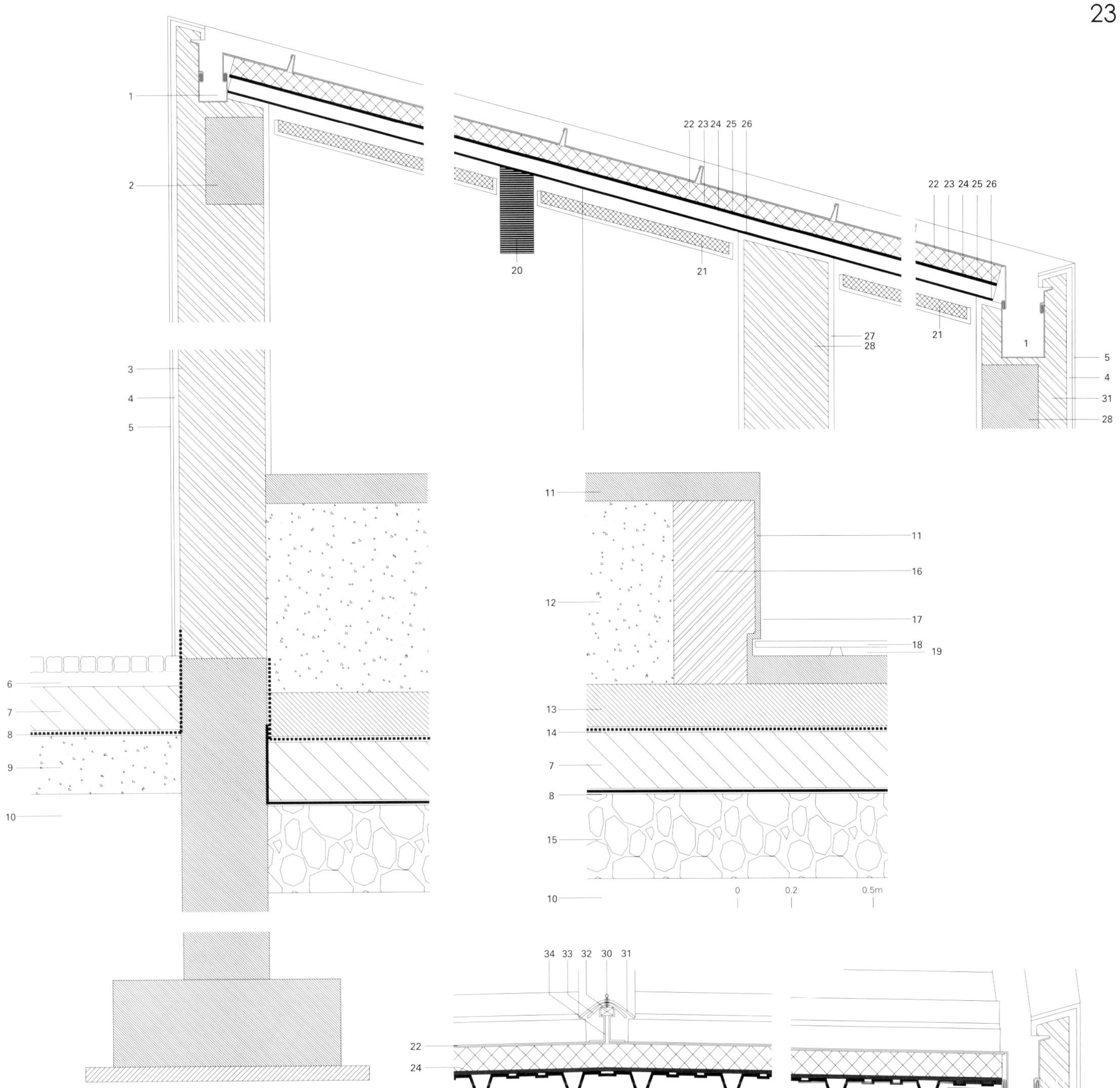

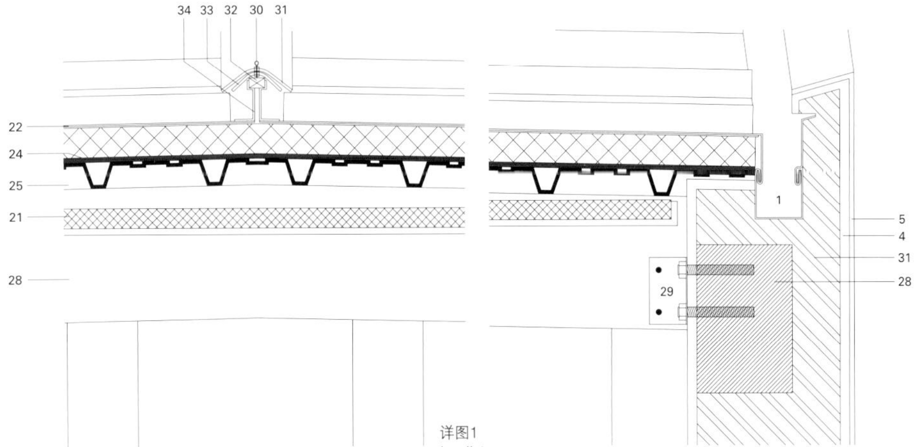

详图1
detail 1

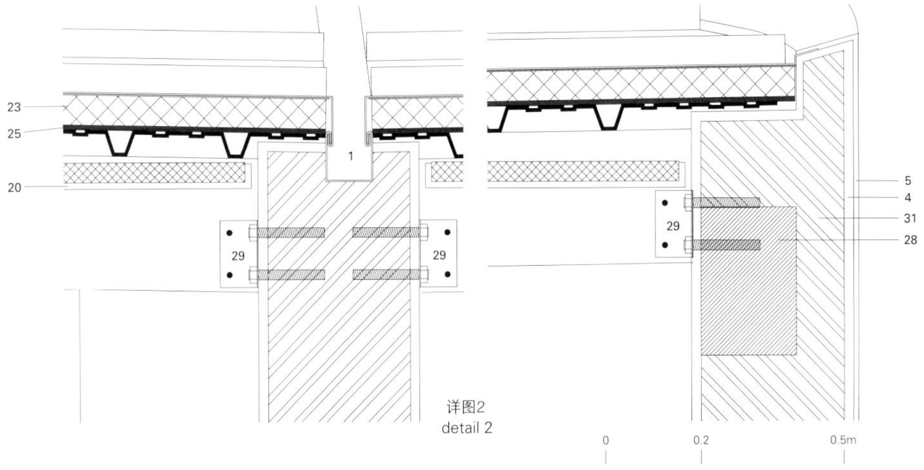

详图2
detail 2

1. flashings and gutter in zinc
2. concrete beam
3. masonry
4. plaster
5. kinky finish
6. basis for finishing exterior floor
7. tout-venant
8. geotextile
9. gravel
10. compacted ground
11. screed stroked
12. filling in lightweight concrete LECA
13. rigging of reinforced concrete
14. PVC membrane
15. rock aggregate
16. concrete block
17. timber footer
18. wood floor
19. wood slate
20. plaster board
21. acoustic insulation
22. zinc roof
23. thermal insulation roof
24. vapor barrier
25. plywood
26. profiled sheet metal
27. wooden beam
28. reinforced concrete beam
29. mounting hardware for the wood beams "simpson"
30. galvanized security support
31. metal easel
32. galvanized metal easel support
33. waterproofing
34. metal bracket and fixing the bridge

项目名称：Saint Anne's Chapel
建筑师：e|348 Arquitectura
甲方：Fábrica da Igreja de Canedo (Canedo's Church)
地点：Sousanil, Canedo, Santa Maria da Feira, Portugal
基地面积：100m²
竣工时间：2009.7
造价：EUR 140,000
摄影师：©FG+SG Fotografía de Arquitectura

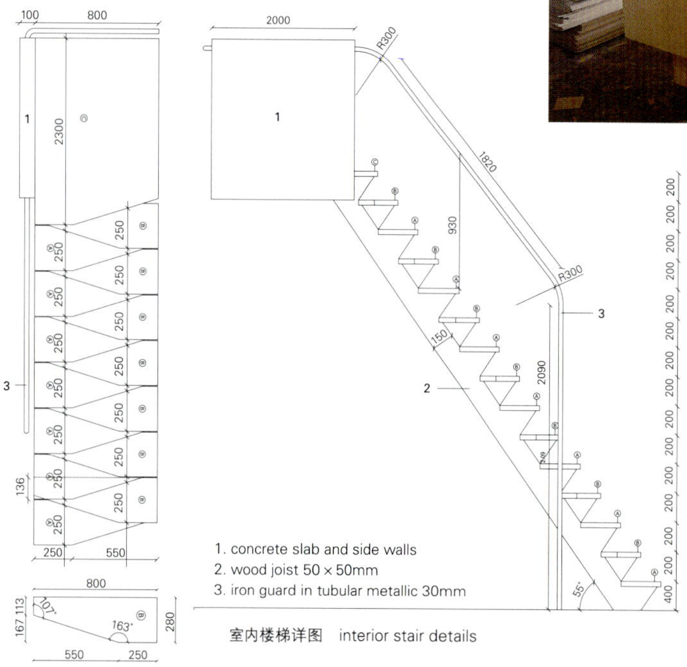


室内楼梯详图　interior stair details

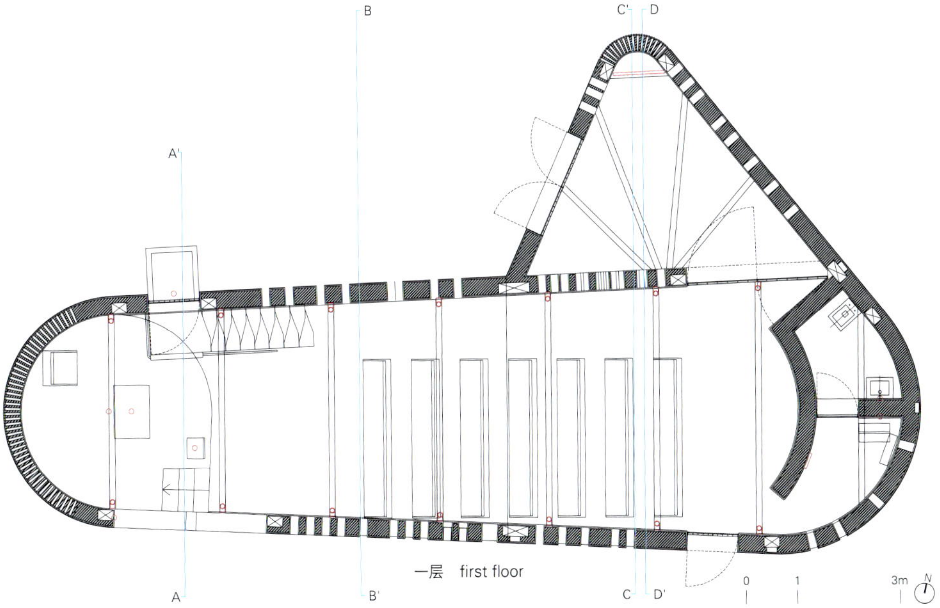


一层　first floor

# 立方体住宅

Shinichi Ogawa & Associates / urbanist architect

项目名称：Cube House
建筑师：Shinichi Ogawa & Associates / urbanist architect
地点：Kanagawa, Japan
基地面积：120.80m²
建筑面积：72.25m²
总建筑面积：114.14m²
竣工时间：2009
摄影师：©Nacasa & Partners Inc.

平面图01
plan 01

平面图02
plan 02

平面图03
plan 03

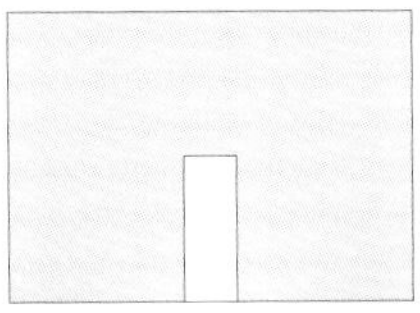
立面
elevation

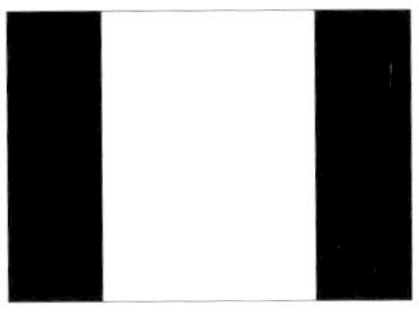
剖面图
section

该立方体住宅位于日本神奈川县川崎市新住宅区的一角，在XYZ三个方向上各由一组1.5m的格子组成。

住宅整体呈9m×9m×6m的立方体结构，内部有4.5m×4.5m×6.0m的中空空间。大厅、小孩卧室和厨房都环绕着一层的客厅和餐厅设计。在二层，卧室、步入式衣橱、浴室、庭院都围绕着双层高6m的天井布置。室内的每个房间都通过客厅的中空空间连在一起。这个中空空间也割开了天花板，让日光能照入到室内。在住宅内，所有的活动、季节的变化、时间的流逝和人的行为都是通过这个中空空间展示出来的。

## Cube House

The house is built on a 1.5m grid module in all XYZ directions. This site is at a corner of the new residential area in Kawasaki-shi, Kanagawa, Japan.

The space is composed of a solid cube of 9m x 9m x 6m and a cubic void of 4.5m x 4.5m x 6.0m. The hall, children's rooms, and the kitchen are laid out to surround the living-dining room on the first floor.

On the second floor, the bedroom, walk-in closet, bathroom, courtyard, are laid out to surround the double floor-height well which is 6.0m. Each room of the interior space is connected through the void of the living room. And the void cuts off the sky as a geometric form, letting the sunlight inside. In this house, all events, the changes of the seasons, course of time, and human activity, are created through the void.

Shinichi Ogawa & Associates/urbanist architect

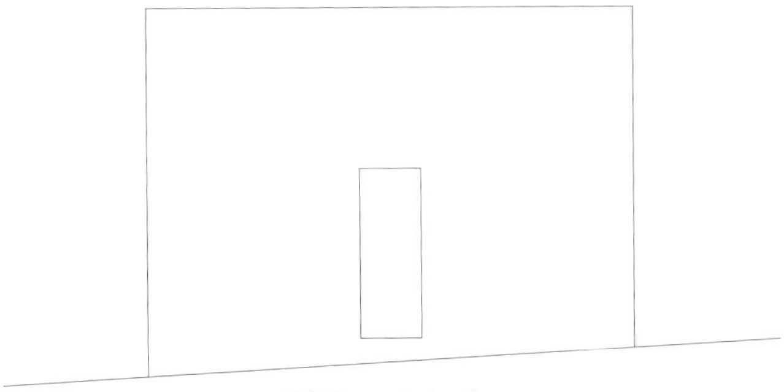
南立面　south elevation

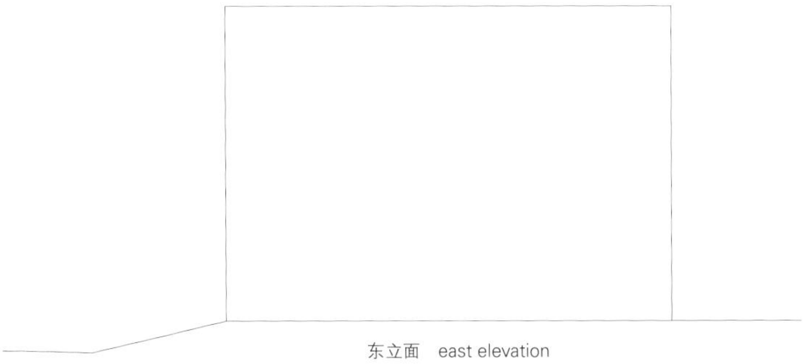
东立面　east elevation

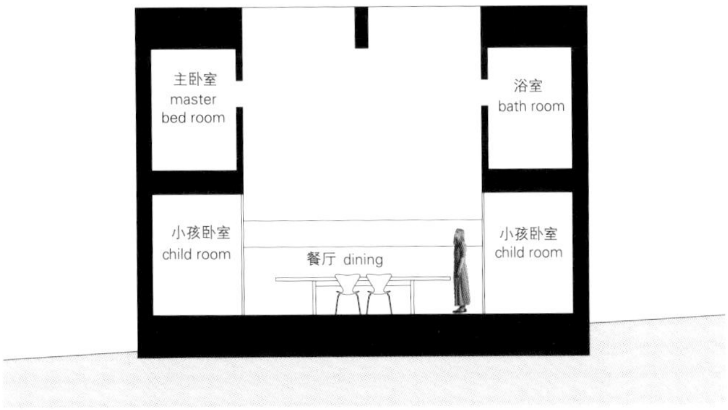

A-A'剖面图　section A-A'

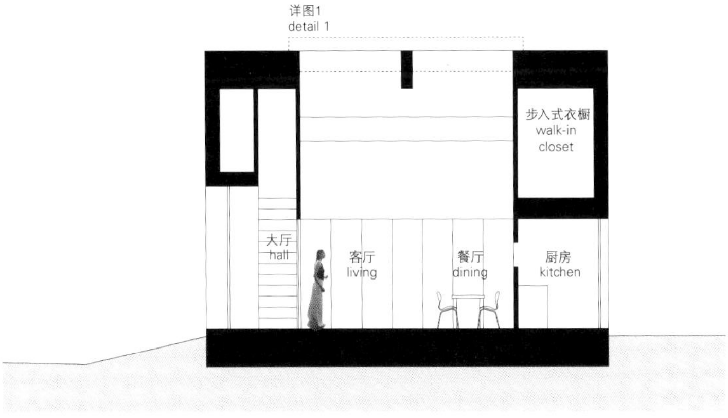

B-B'剖面图　section B-B'

0　2　5m

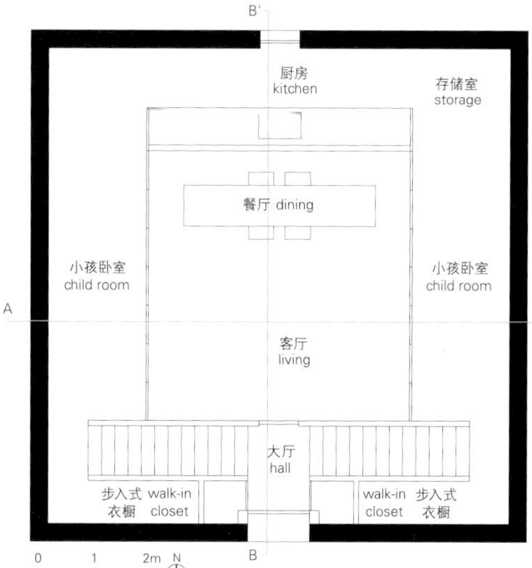


一层 first floor

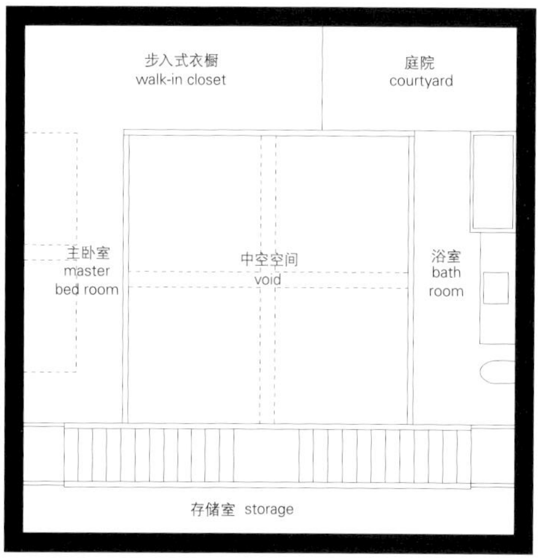


二层 second floor

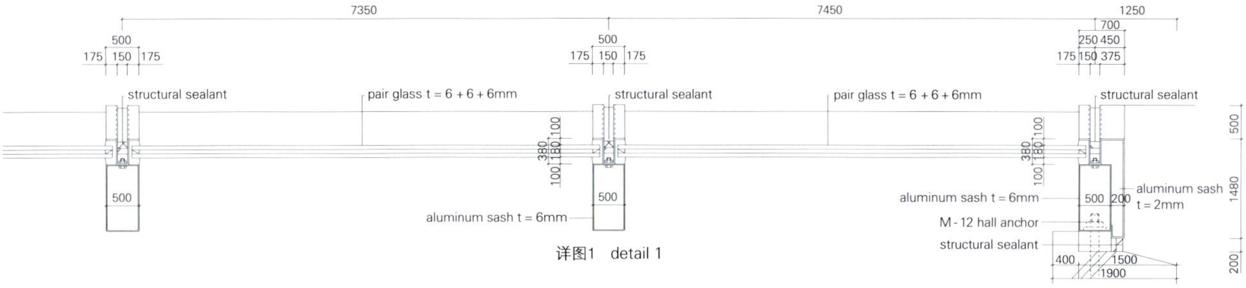
7350
7450
1250
500
175 150 175
700
250 450
175 150 375
structural sealant
pair glass t = 6 + 6 + 6mm
structural sealant
pair glass t = 6 + 6 + 6mm
structural sealant
380
100 180 100
500
aluminum sash t = 6mm
aluminum sash t = 6mm
M - 12 hall anchor
structural sealant
aluminum sash
t = 2mm
500 200
400
1500
1900
500
1480
200

详图1 detail 1

# Aso展廊
Architecture H

我们遇到了一对令人称奇的夫妇，两人都极富激情：一个是林业学教授，另一个是他的妻子，她在自己的院子里培植不同寻常的美丽野花和原生植物已长达30年。拥有一间放置自己关于野生植物的艺术作品的小小展廊一直是他们的梦想。唯一可用的地方是庭院（$90m^2$）的一个小角落，那里有间老花房。

整个项目令人耳目一新：一个花的世界。我们不仅需要处理空间狭小的问题，同时还需弄清楚怎样去实现关于野花展廊的设想。展廊能为忙碌的现代生活提供梦幻时光及其所需的宁静吗？

从一开始，我们就想设计一个可以让人沉思、远离日常生活的沉静空间。由于混凝土质地粗糙，给人感觉沉静，因此我们选择它来衬托花的精致。

从外面看，整栋建筑不过是一个立面上有着大型滑动木门的混凝土盒子。打开大门，一条黑暗的隧道般的坡道将人引向展廊。走廊尽头，人们可以进入到内庭院（$10m^2$）开放的空间里，看见里面的水池。内庭院是整座建筑的中心，将内部空间分成主展室和平台，二者通过内庭院遥相对应。在整个空间内，光与风自由地在水池上掠过，当游客的目光越过宁静的水面看向花朵时，思绪会超越肉眼所看到的内庭景象。

在整个项目过程中，我们和甲方都有着难忘的愉快体验。很希望某一天我能在这里举办一场古琴音乐会。飘过水池的声音就像来自你内心深处的仙乐，会有一种不真实的美。

## Gallery Aso

We met an amazing couple whose passions seemed entwined: a professor of forestry and his wife who, for 30 years, had grown unusual and beautiful wildflowers and native plants in her courtyard. Owning a small gallery for her artistic work with wild plants had long been their dream. The only possible site was a small corner of the courtyard (90 square-meter), where an old greenhouse stood.

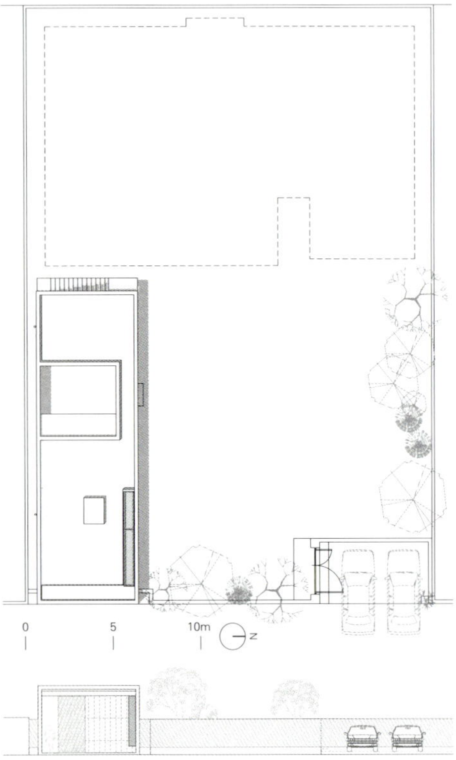

南立面　south elevation

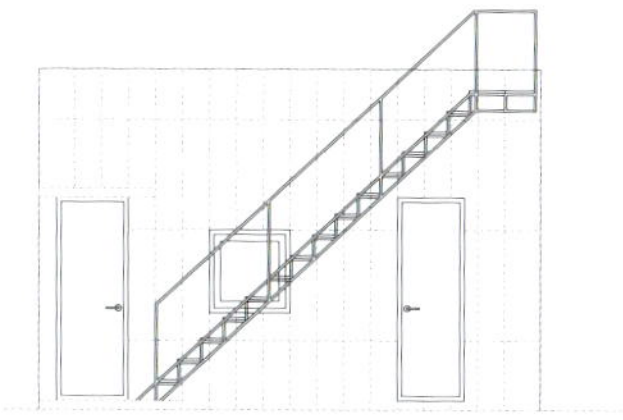

西立面　west elevation

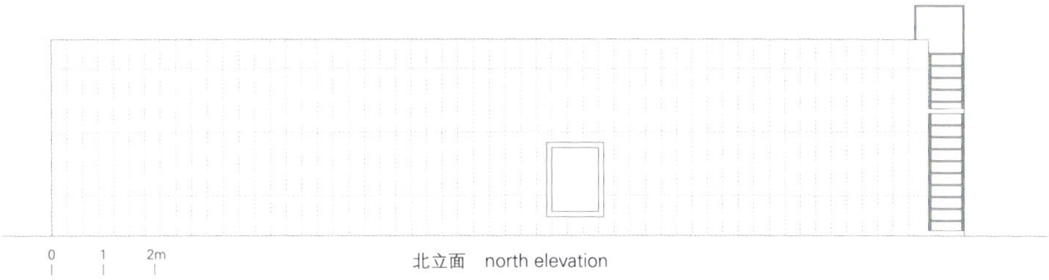


北立面　north elevation

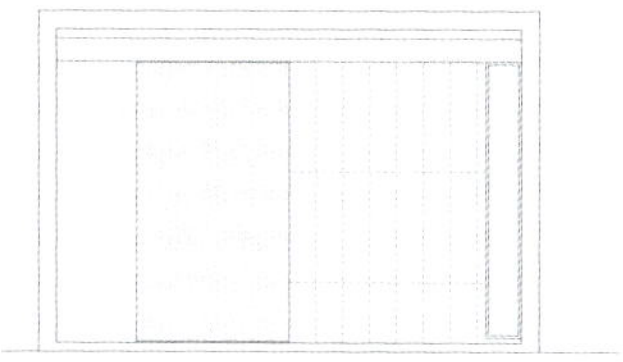

东立面　east elevation

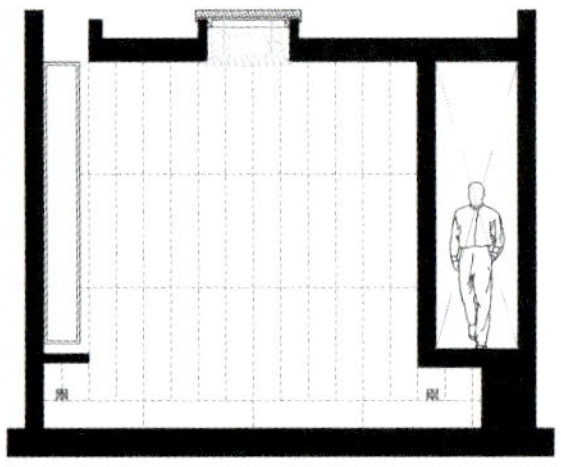

A-A'剖面图　section A-A'

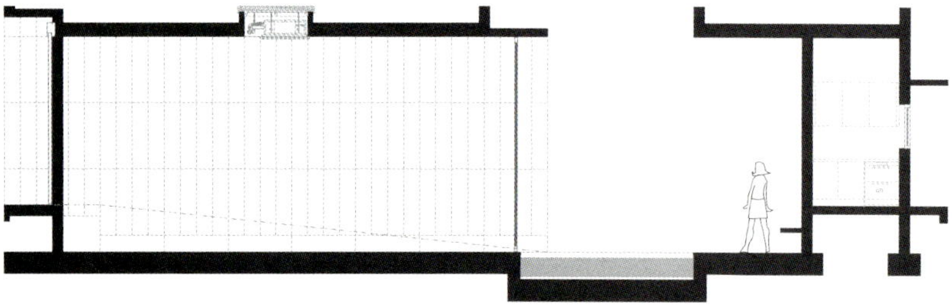

B-B'剖面图　section B-B'

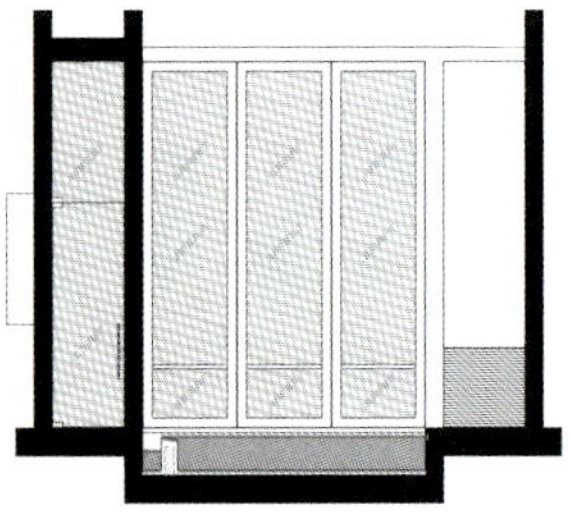

C-C'剖面图　section C-C'

D-D'剖面图　section D-D'

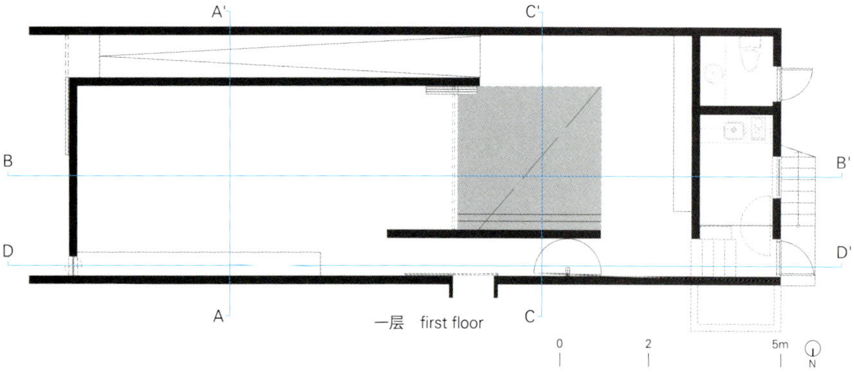


一层　first floor

The program was quite refreshing: a place for contemplating flowers. We needed not only to deal with the smallness but also figure out how to conceptualize a gallery space for wildflowers. Could the gallery be both a dramatic moment in this hectic contemporary life and provide the needed serenity?

From the very beginning, we wanted the gallery space to provide sanctuary for contemplation and to transcend everyday life. Concrete, with its brutal texture and silence, would be a potent counterpoint to accentuate the delicacy of the flowers.

From outside, the building is nothing more than a concrete box with a large wooden sliding door on the facade. When it is opened, a dark, almost tunnel-like ramp leads visitors into the gallery. At the end of the corridor, visitors encounter the openness of the inner patio (10 square-meter) and its water pool. This inner patio is the core of the building that divides the interior into 3 bays: the main exhibition room and the platform facing each other across the patio. Throughout the void, light and wind flow freely over the pond and, when a visitor's eyes go out towards the flowers across this stillness of water, one's mind looks beyond what it is seen to the inner landscape of oneself.

During all the processes, this project was delightful and unforgettable for both of the clients and us. Someday I'd love to organize a koto concert here. The sound from beyond the water would be surreal and beautiful, just like the one coming from your inner mind. JiWon Lee

项目名称：Gallery Aso
建筑师：HyunJai Lee + Ji Won Lee
项目团队：Architecture H
甲方：DeokSoon Cho, JongYun Lee
地点：Daegu, South Korea
用途：gallery
基地面积：716.70$m^2$
建筑面积：187.36$m^2$(existing) + 73.88$m^2$(annex)
总建筑面积：238.40$m^2$
建筑覆盖率：36.45%
总建筑比率：33.26%
竣工时间：2008
摄影师：courtesy of the architect
(©EuiRyong Hwang)

# 日落小礼拜堂

BNKR Arquitectura

该建筑的宗教任务首先是希望这里作为新婚夫妇新生活开始的地方，第二个目的刚好相反，希望在这里寄托对所爱的人的思念。两个天敌般对立的功能的委托似乎成了这个设计的主要驱动力。前者赞美生命，后者哀悼死亡。对立的要素开始碰撞：玻璃与混凝土、透明度与坚实度、空灵与沉重、古典的比例与明显的混沌、脆弱与坚不可摧、短暂与持久。

甲方的要求直白简单：首先要充分利用小礼拜堂外壮丽的景色；其次，太阳须精确地从圣坛十字架后方落下（当然，这样的情况只有在昼夜平分的时候才会出现，一年之中只有两次）；最后一个要求同样很重要，希望建筑融入场地。

两个主要的因素遮挡了观看日落的视线：大型的树木和丰富的植被以及庞大的巨石。为了避开这些不利因素（从伦理、精神、环境、经济各方面原因综合考虑，搬开大石头是不可能的），小礼拜堂至少需要升高5m。

这里是一片充满异域风情、美景如画的处女地，因此我们努力减少了近一半的接地面积，以降低对场地的影响。

阿卡普尔科山由巨大的花岗岩堆积而成，我们力求将教堂处理得如同山顶上的“另一块巨石”。

**Sunset Chapel**

Our first religious commission was a wedding chapel conceived to celebrate the first day of a couple's new life. Our second religious commission had a diametrically opposite purpose: to mourn the passing of loved ones. This premise was the main driving force behind the design; the two had to be complete opposites; they were natural antagonists. While the former praised life, the latter grieved death. Through this game of contrasts all the decisions were made: glass vs. concrete, transparency vs. solidity, ethereal vs. heavy, classical proportions vs. apparent chaos, vulnerable vs. indestructible, ephemeral vs. lasting.

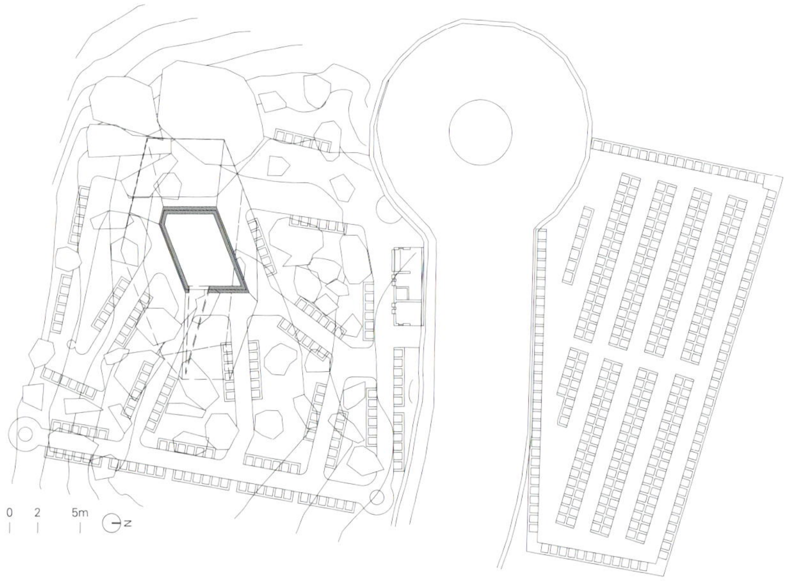
0
2
5m
N

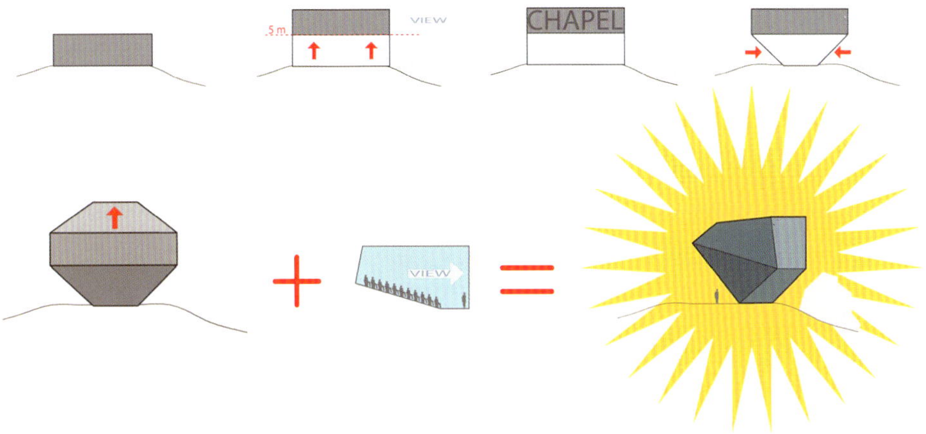


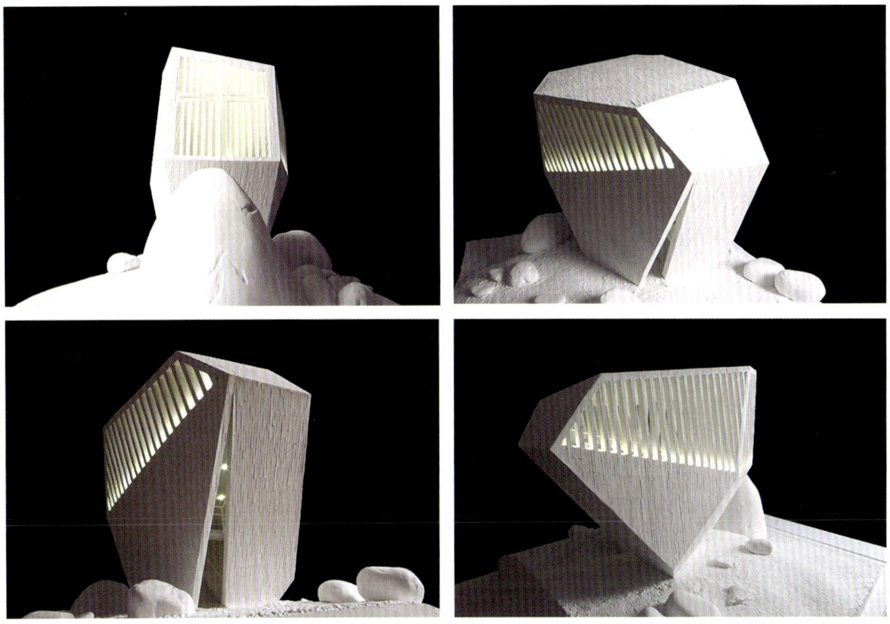

项目名称：Sunset Chapel
建筑师：BNKR Arquitectura
项目负责人：Mario Gottfried, Javier González, Roberto Ampudia
项目团队：Mario Gottfried, Rodrigo Gil, Roberto Ampudia, Javier González, Óscar Flores, David Sánchez, Diego Eumir, Guillermo Bastian, Adrian Aguilar
合作者：Jorge Arteaga, Zaida Montañana
结构工程师：Juan Felipe Heredia, José Ignacio Báez
机电管道工程师：SEI
照明：Ricardo Noriega (Noriega Iluminadores)
施工：Fermín Espinosa & Francisco Villeda (Factor Eficiencia)
地点：Acapulco, Guerrero, Mexico
建筑面积：120m$^2$
竣工时间：2011.2
摄影师：courtesy of the architect (©Esteban Suárez)

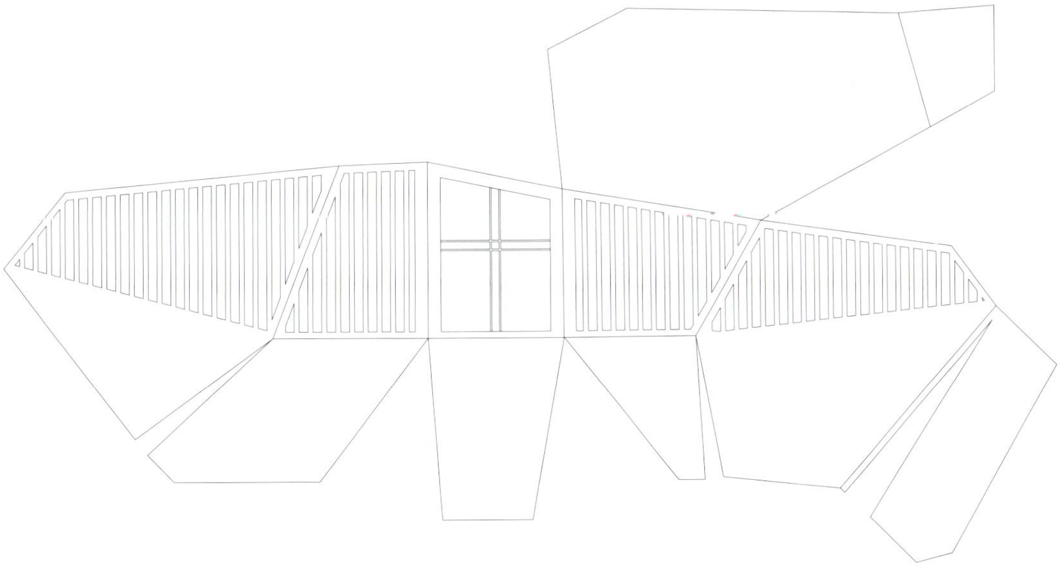

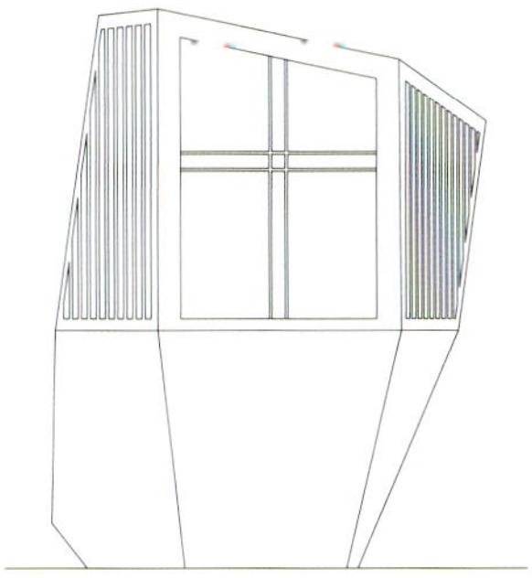

西立面　west elevation

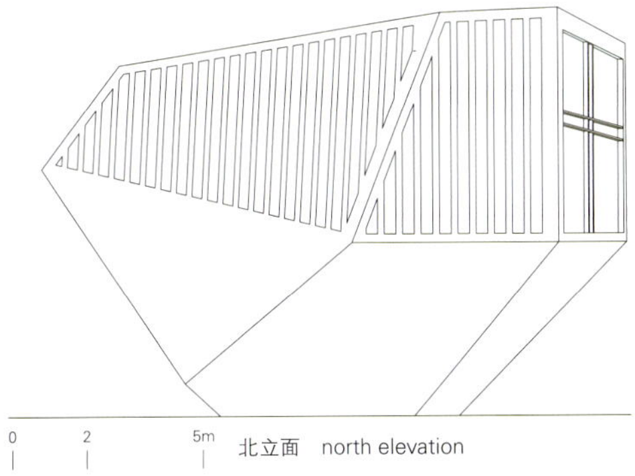


北立面　north elevation

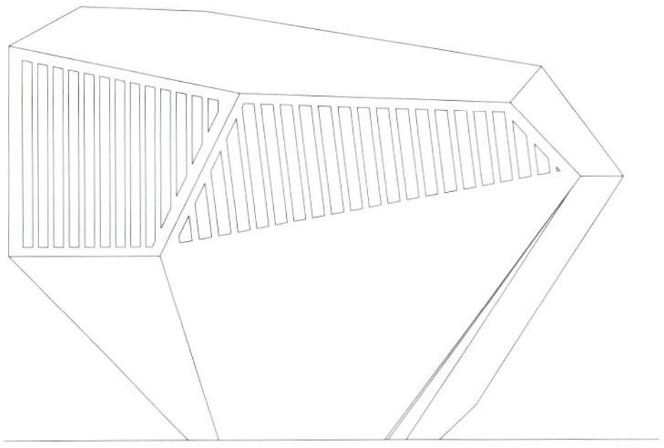

南立面　south elevation

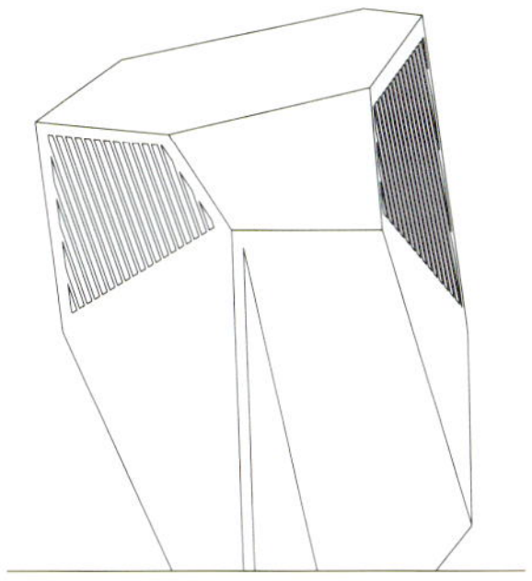

东立面　east elevation

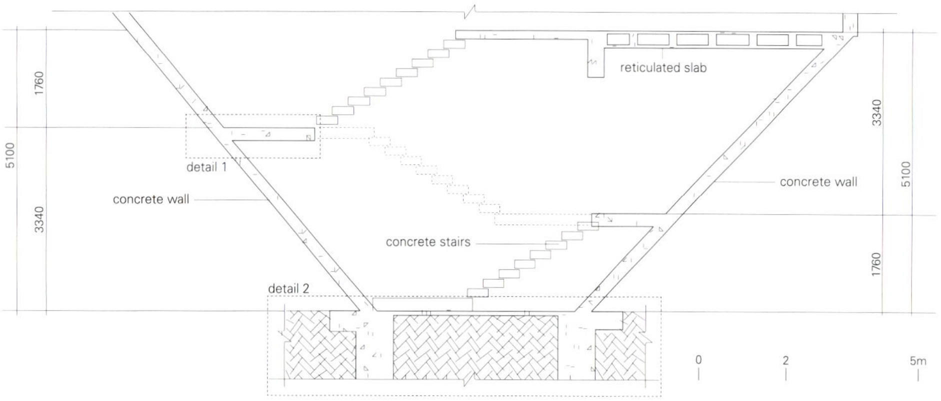

楼梯立面　stair elevation

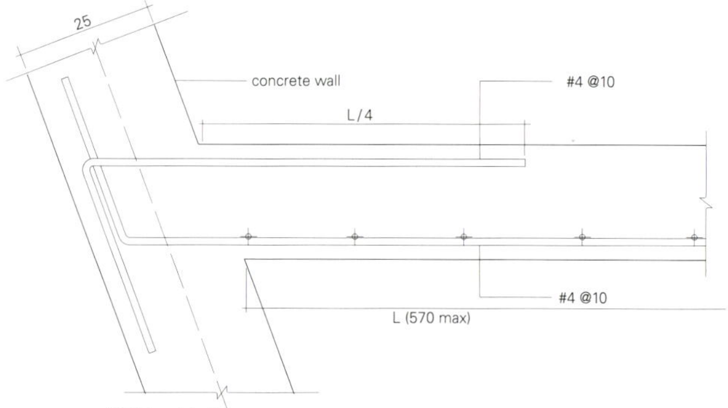

详图1　detail 1

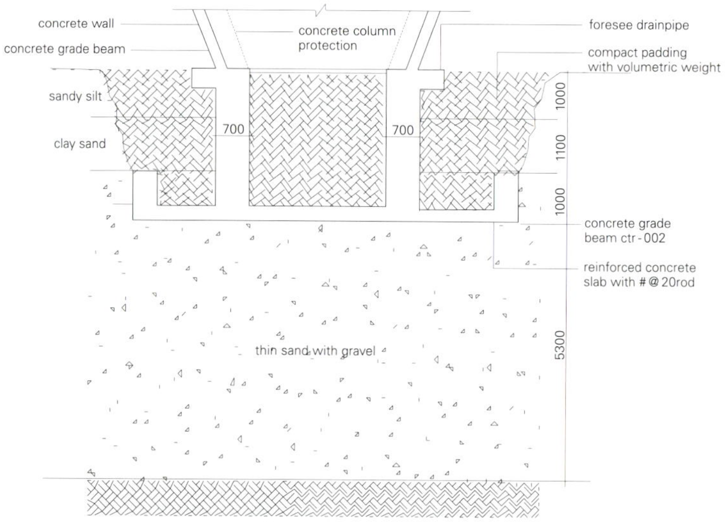

详图2　detail 2

Y MEDITACION
A DIOS.

The client brief was pretty simple, almost naive: first, the chapel had to take full advantage of the spectacular views. Second, the sun had to set exactly behind the altar cross (of course, this is only possible twice a year at the equinoxes). And last but not least, a section with the first phase of crypts had to be included outside and around the chapel.

Two elements obstructed the principal views: large trees and abundant vegetation, and a behemoth of a boulder blocking the main sight of the sunset. In order to clear these obstructions (blowing up the gigantic rock was absolutely out of the question for ethical, spiritual, environmental and, yes, economical reasons) the level of the chapel had to be raised at least five meters.

Since only exotic and picturesque vegetation surrounds this virgin oasis, we strove to make the least possible impact on the site, reducing the footprint of the building to nearly half the floor area of the upper level.

Acapulco's hills are made up of huge granite rocks piled on top of each other. In a purely mimetic endeavor, we worked hard to make the chapel look like "just another" colossal boulder atop the mountain. BNKR Arquitectura

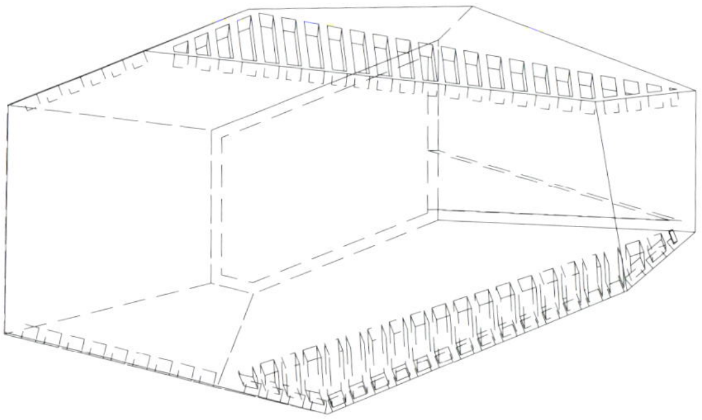

屋顶层 roof floor

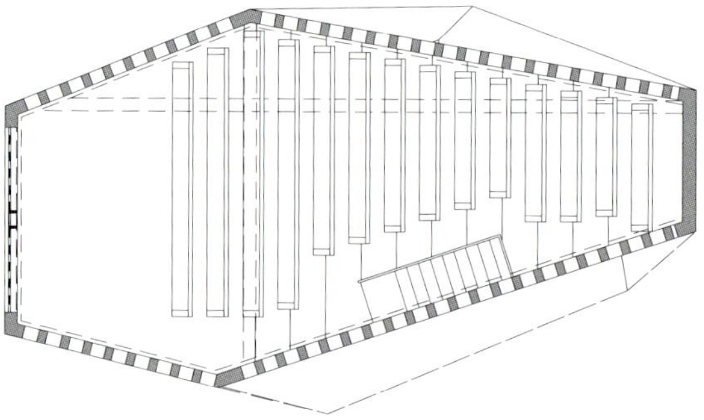

三层 third floor

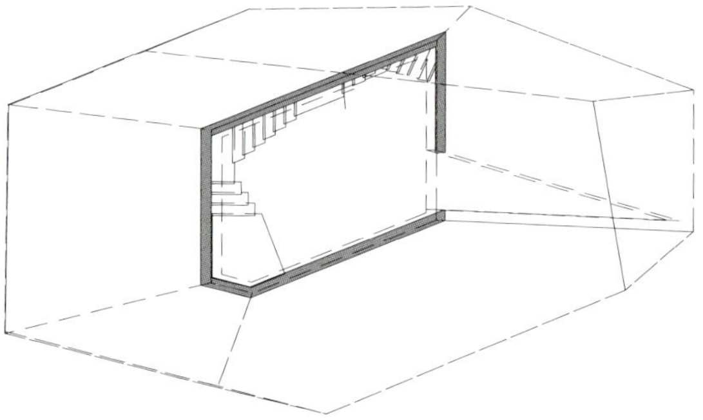

二层 second floor

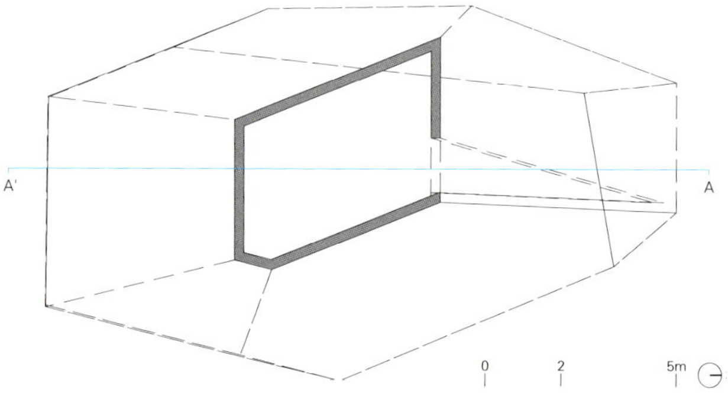


一层 first floor

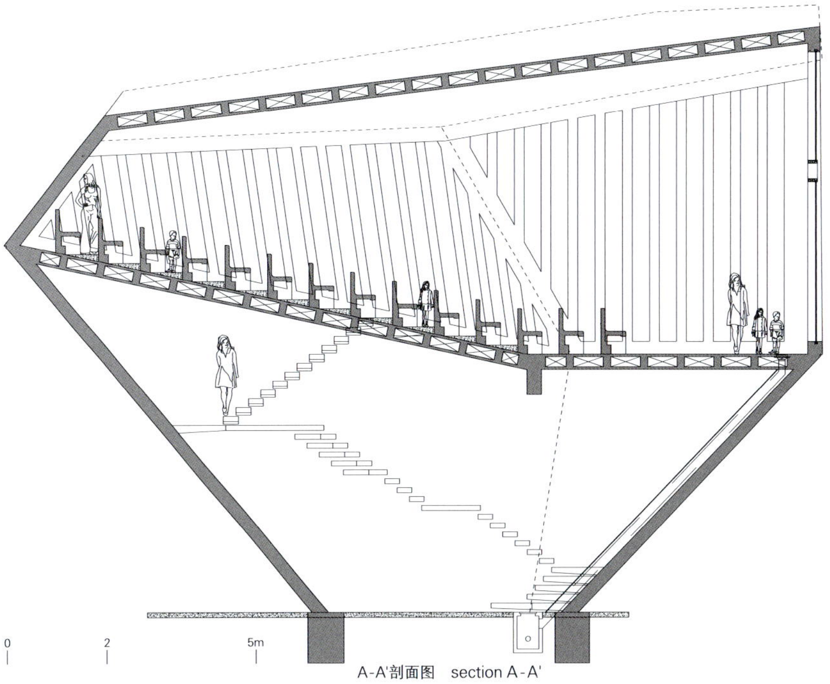


A-A'剖面图　section A-A'

# 立方体

5468796 Architecture

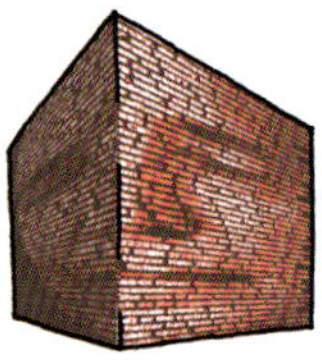

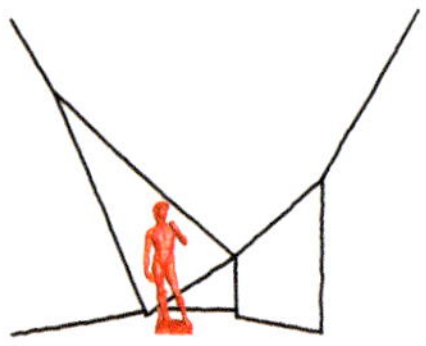

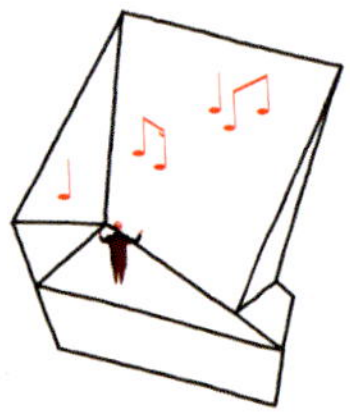

“立方体”是一个露天的表演场馆，建于温尼伯交易区的老仓库对面。在项目最后移交的阶段，舞台成为一个多功能的环境，可以从一个充满活力的表演空间转变为一个短暂的、发光的、互动的展馆和焦点。

舞台的外壳被串在一起的菱形挤压型材组成的薄膜包裹，形成灵动的幕布，幕布抬升之后可以看见下面的舞台和内部的结构。继而，可伸缩的外壳变成了卷起的、波浪形的天花板，该天花板成了舞台背景，并可根据舞台的音响效果进行调整。外壳关闭后，菱形块把光线或者影像捕捉并折射到菱形块的表面，形成独特的像素矩阵。经过编程的灯光投射在像素化的外表皮上，为交互式地展现不同季节的活动提供了背景，并为当地艺术家的积极活动提供便利。舞台关闭的时候，其内部适合举行小型聚会和展览。该项目一个重要的目标是提供一面安全遮屏，该安全遮屏可根据各种节目开放或者关闭。我们面临的挑战是创造独特的网格——既足够柔软，可以方便地悬垂放下，又可作为有足够刚性的坚实遮屏。早期的基本结构演变为从未有过的锁子甲形状——一种柔韧的保护薄膜，该薄膜可以形成场馆的外罩。

第二个关键目标是探索薄膜捕捉投射到其表面上的影像的能力。遮屏到舞台后墙（这里安装有投影机）的距离较小，影像将会映在遮屏背面，但在人们看向舞台时却可以观察到，这些都是技术使用的限制因素。我们研究了有效捕捉投影图像必需的元素的尺寸和数目。研究完成后可以确定在短距离的投影路径内我们需要哪些“短焦”投影仪来扩展图像。此外，我们通过这种原型来探索影像单元的深度、投影能力、分辨率和图像的完整性。

我们与当地制造商一起合作，委托他们制造出不同挤压型材的样品，最初是用小铝板焊接而成的。确定了合适的原型后，我们便向制造商定制挤压铝型材，这些铝板可以切割成一块块的小铝板。单块的挤压铝型材上开有小洞，用航空钢丝绳串起。每一块的朝向可以在沿钢丝绳的方向上进行上下交替调整，每三块相邻的小铝板铆接在一起。因此，外壳薄膜在保持连续表面的完整性的同时，又能很好地串连在一起。预先串起来的遮屏板在铆接之前在现场组装起来，形成一个连续的装配体。

最终安装时，遮屏被垂直悬挂起来，组成8.5m×8.5m×8.5m的立方体。遮屏抬升之后会折叠成波浪形，露出下面的舞台空间，形成一个音响背景。遮屏关闭时，抛光的挤压铝型材形成一个光反射表面，与地面的灯光和投影仪一起形成动态的、按设计变化的灯光效果。

对全年舞台节目的讨论激发了设计团队的积极性，把功能的组成部分（安全、遮屏、帐篷）发展成了一个新的项目。

因此，外表皮不仅起到保护罩的作用，而且还是一个全新的标志性建筑。

## The Cube

"The Cube" is an open-air performance venue set against the backdrop of historic warehouses in Winnipeg's Exchange District. In its final rendition the stage is articulated as a multi-functional environment that shifts from a vibrant performance space to an ephemeral, glowing, interactive pavilion and focal point.

The outer shell of the stage is a dynamic membrane composed of diamond extrusions strung together to form a flexible curtain that draws back to reveal stage and structure within. The retracted skin in turn becomes a draped and undulating ceiling landscape providing a backdrop for performance and allowing adjustments to the stage's acoustics. When closed, the angled diamond pieces capture and refract light or images to their outer surface, creating a unique pixel matrix. Programmable lighting shines onto the pixelated skin offering a canvas for interactive displays seasonally programmed and available for active engagements by local artists. The stage in its closed position also accommodates small gatherings and exhibitions inside. An important objective for the

项目名称：OMS Stage
建筑师：5468796 Architecture
项目团队：Sharon Ackerman, Mandy Aldcorn
Ken Borton, Aynslee Hurdal, Michelle Heath, Johanna Hurme
Cristina Ionescu, Grant Labossiere, Colin Neufeld, Zach Pauls
Sasa Radulovic, Shannon Wiebe
结构：Lavergne Draward & Associates
照明设计：Ambiances Lighting + Visual Design
项目经理：Mark Penner(Greenseed Development)
金属制造商：Melvin Kleinsasser(Central Prairie Products)
地点：Winnipeg, Manitoba, Canada
项目面积：784m$^2$
材料：cast in place concrete agilia, custom aluminum extrusions
竣工时间：2010

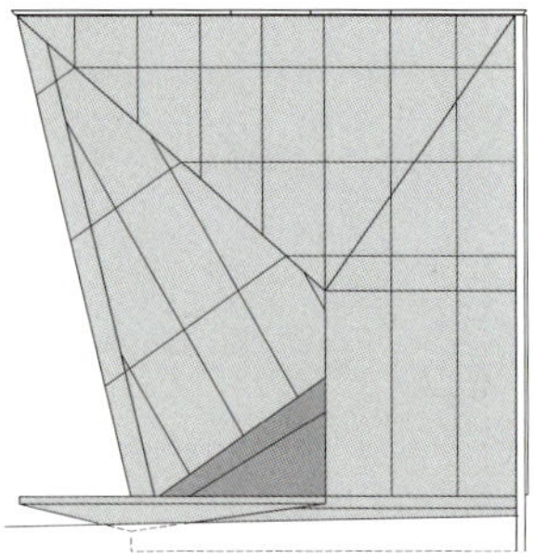

东立面（无遮屏）
east elevation (without screen)

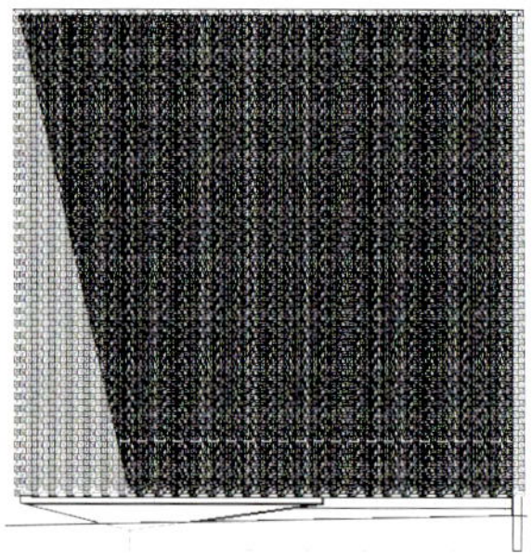

东立面（有遮屏）
east elevation (with screen)

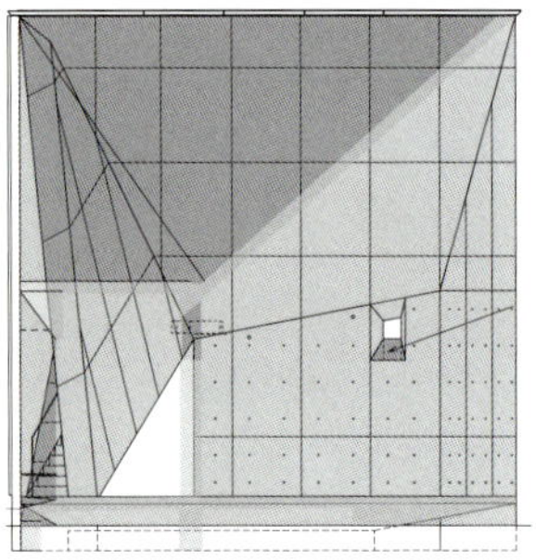

南立面（无遮屏）
south elevation (without screen)

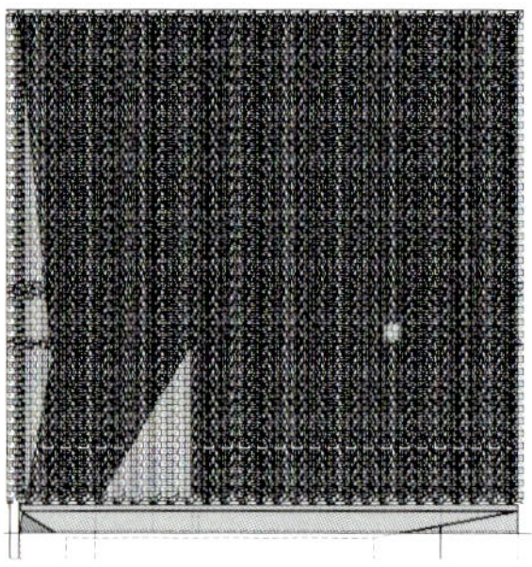

南立面（有遮屏）
south elevation (with screen)

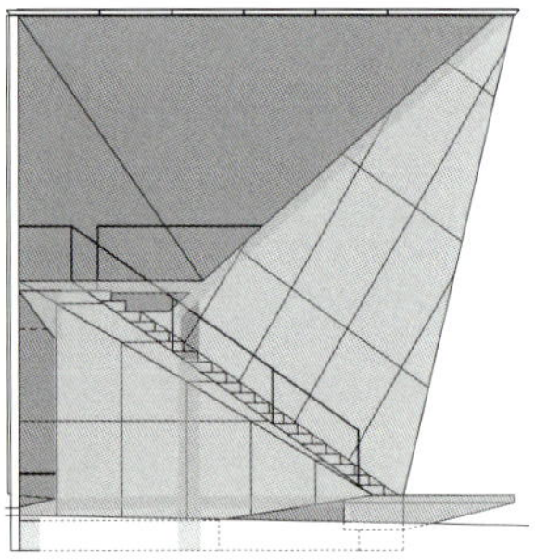

西立面（无遮屏）
west elevation (without screen)

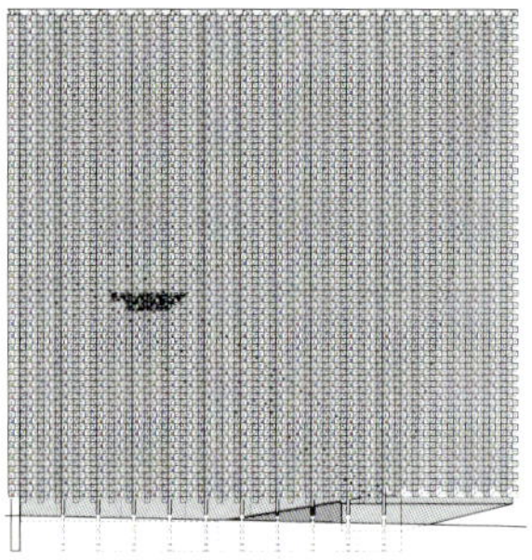

西立面（有遮屏）
west elevation (with screen)

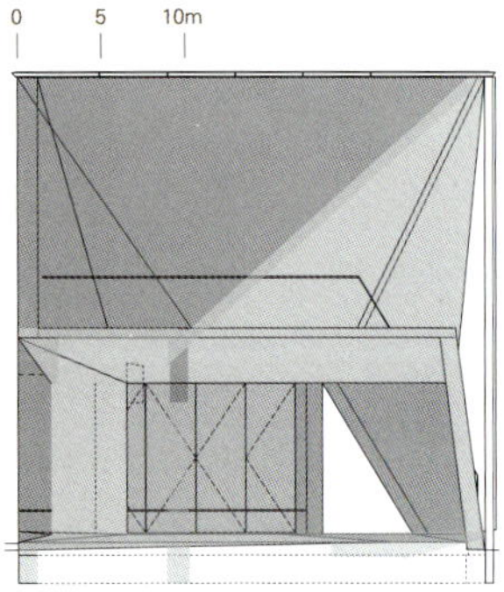


北立面（无遮屏）
north elevation (without screen)

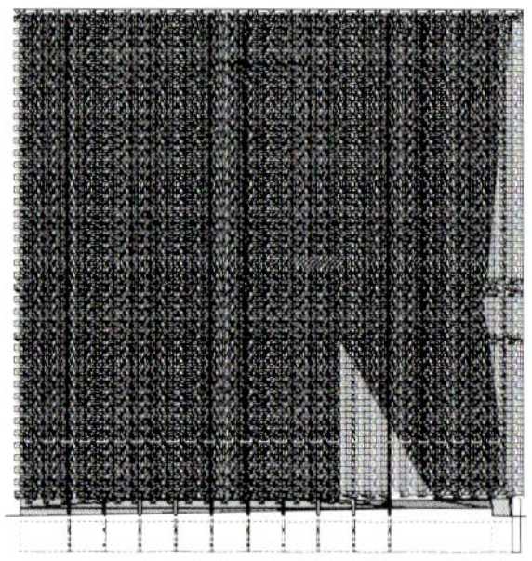

北立面（有遮屏）
north elevation (with screen)

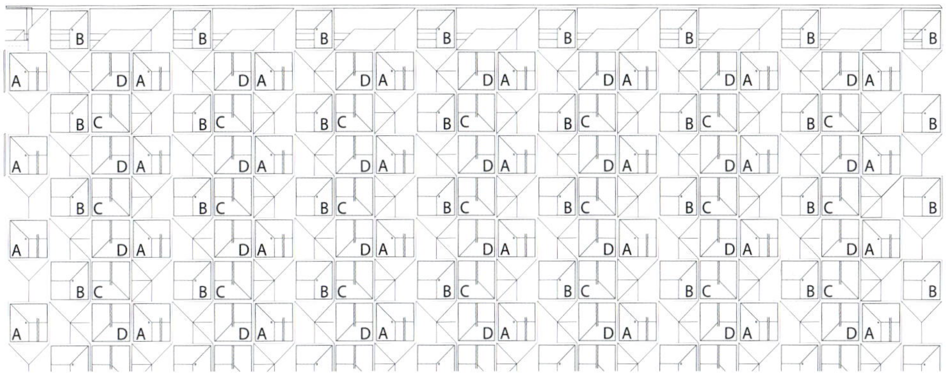
A
B
C
D

在一座郊区住宅的后院里有一个由铜和玻璃构成的悬挂结构，会让人联想到日式灯笼，事实上它是一个茶室，一个能够让人静心思考的地方，也是家庭音乐表演的舞台。这个“灯笼”仿佛是一颗漂浮的宝石，穿过一排排能净化心灵的竹子之间的空间，人们就可以进入到它的内部。

游客登上一段折纸楼梯后，就会看到这座建筑最后的自然元素：一扇10cm厚的不透明木门。站在这里，游客仿佛是一个表演者，似乎能够冥想来世。

构建这个项目时我们遇到的挑战就是：既要完成茶室的建造，又要把其对周围环境的影响降到最低。整个茶室预先在工厂里建造好，然后被运到这里组装起来。整个结构与地面接触的面积很小。

两个U形钢支架固定在地面上，并且牢固相连，从而支撑起整个结构。“悬挂的灯笼”内部的天花板向下倾斜，为茶道营造出一个舒适的空间。天花板和屋顶构成一个正八面体，其中还包括一个三角形框架，用于抵抗外部扭力。悬挂在支架上的钢管穿过屋顶固定在地板上，把整个结构从地面上抬起。这些张拉柱都被置于玻璃墙体内，使整个茶室从外部看起来很轻盈。

入口楼梯由混凝土块和折叠钢板构成。正门由坚固的红木打造而成。茶室外部还有一层铜保护层。茶室地板的材料是花旗松，而天花板是由樱桃木制成的。柱子和室内窗框都是红木的。茶室还有一组红木双开门，打开后可将整个茶室变成一个可供音乐表演的舞台。

**Tea House**

A hanging bronze and glass object inhabits the backyard of a suburban home. The structure, which evokes the image of a Japanese lantern, acts as a tea house, meditation space, and stage for the family's musical recitals. After experiencing the image of the lantern as a singular gem floating in the landscape, one is funneled into a curated procession space between strands of bamboo that is conceived to cleanse the mind and prepare one to enter the object.

After ascending an origami stair, the visitor is confronted with the last natural element: a four inch thick, opaque wood entry door. At this point the visitor occupies the structure as a performer with a sense of otherworldliness meditation.

# 茶室

David Jameson Architect

The challenge of constructing the tea house was to do so while imposing the absolute minimum impact to the site. The entire tea house was prefabricated in a shop then trucked to the site and erected. The structure touches the earth in only a few places, leaving a tiny footprint.

Two U-shaped steel brackets are rooted to the ground and securely attached to each other to create the assembly from which the lantern is supported. The ceiling inside the "hanging lantern" slopes down to create an intimately scaled space for tea ceremonies. The ceiling and roof comprise an octahedron which contains a triangulated framework that is engineered to resist external forces of torsion. Steel tubes, hanging from the brackets, thread down through the roof and attach to the floor to hold the structure off the ground. These tensile columns are placed inside the glass walls to give the exterior of the tea house a visually light appearance.

The entry stair is composed of a concrete mass and a folded steel plate. The entry door is solid mahogany. The exterior of the tea house is clad in bronze. The floor of the tea house is end-block Douglas fir. The ceiling panels are made of cherry wood. The columns are clad in mahogany and the interior window frames are also mahogany. There are a set of mahogany double doors that open up the broad side of the tea house which allows it to become the stage for a music recital. David Jameson Architect

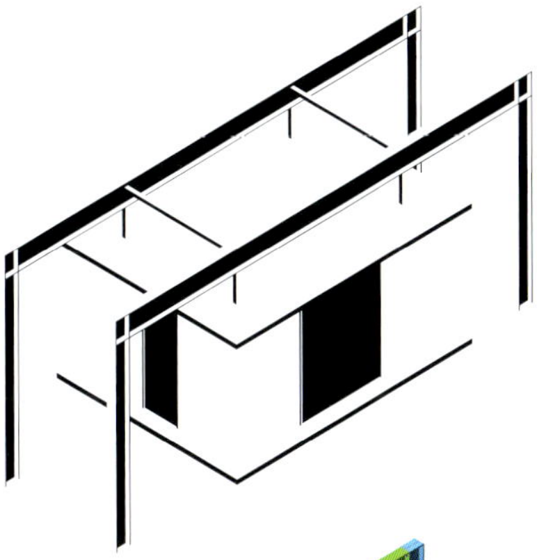

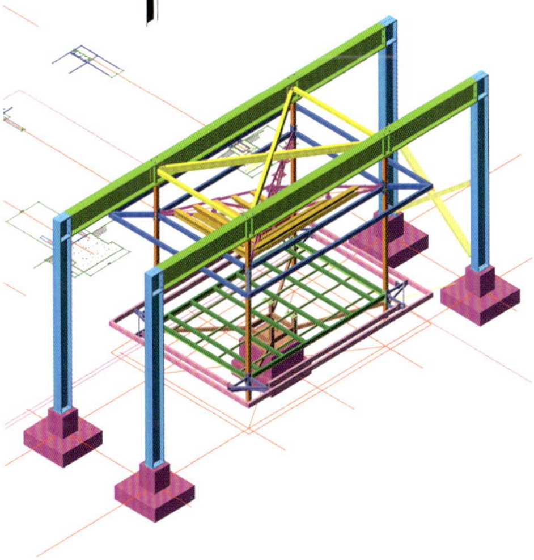

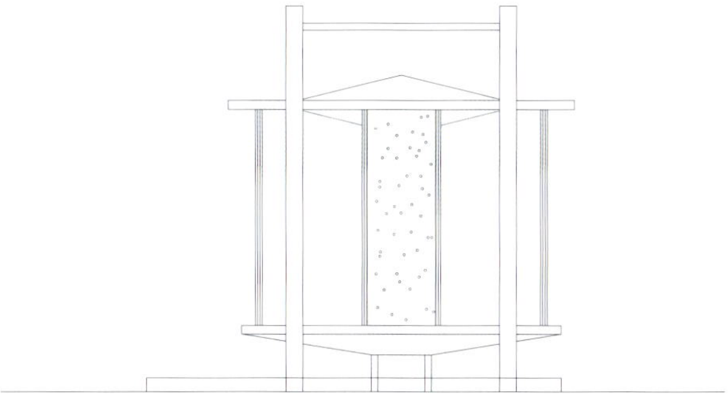

东立面 east elevation

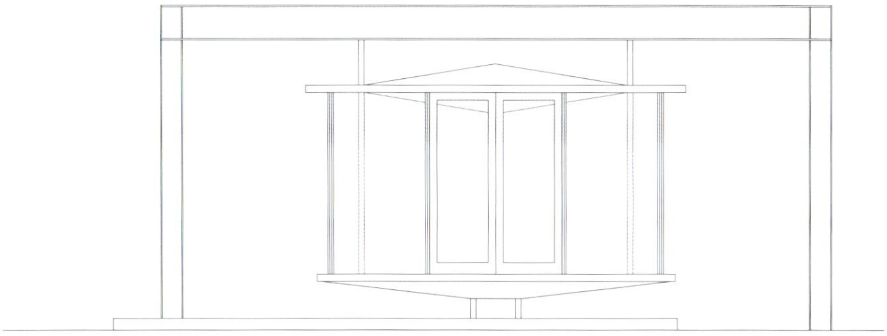

南立面 south elevation

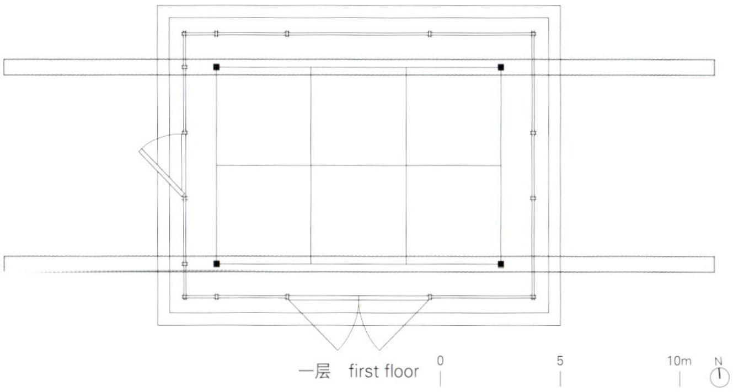


一层 first floor

项目名称：Tea house
建筑师：David Jameson Architect
结构工程师：David Linton
地点：5704 Huntington Parkway, Bethesda, Maryland, USA
基地面积：534.8m²
建筑面积：14.3m²
竣工时间：2009
摄影师：courtesy of the architect (©Paul Warchol)

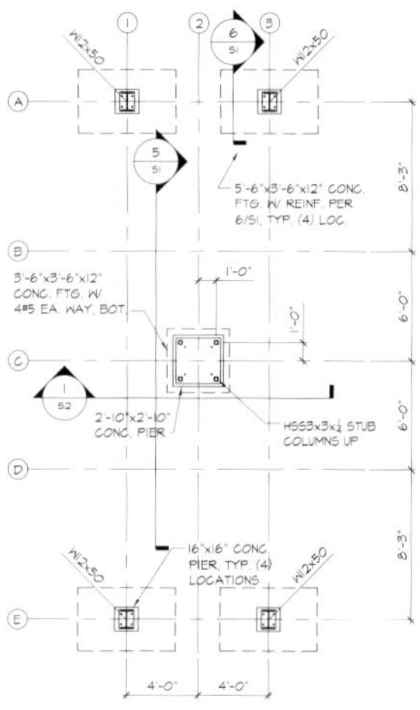


FOUNDATION PLAN
SCALE: 1/4" = 1'-0"

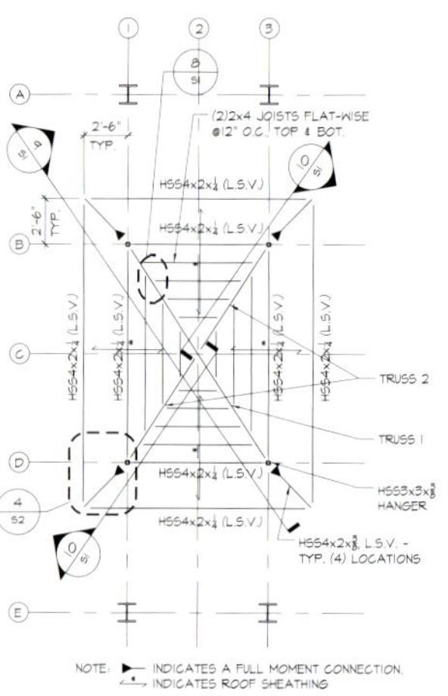


ROOF FRAMING PLAN
SCALE: 1/4" = 1'-0"

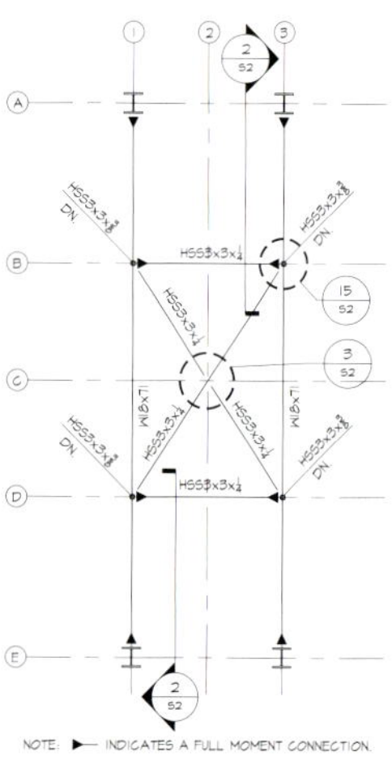


GIRDER FRAMING PLAN
SCALE: 1/4" = 1'-0"

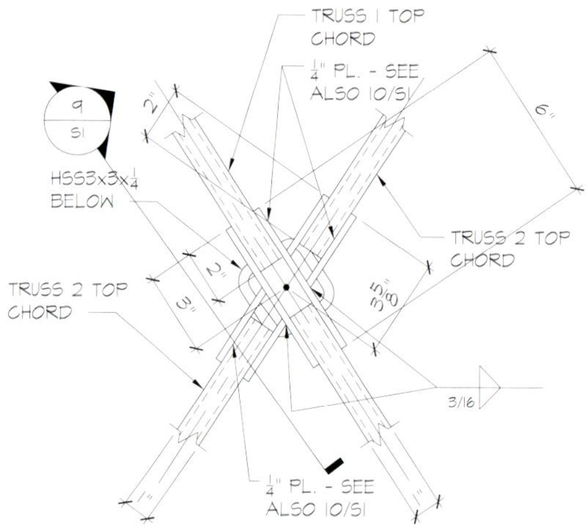


TRUSS CONN. DET.
SCALE: 3" = 1'-0"

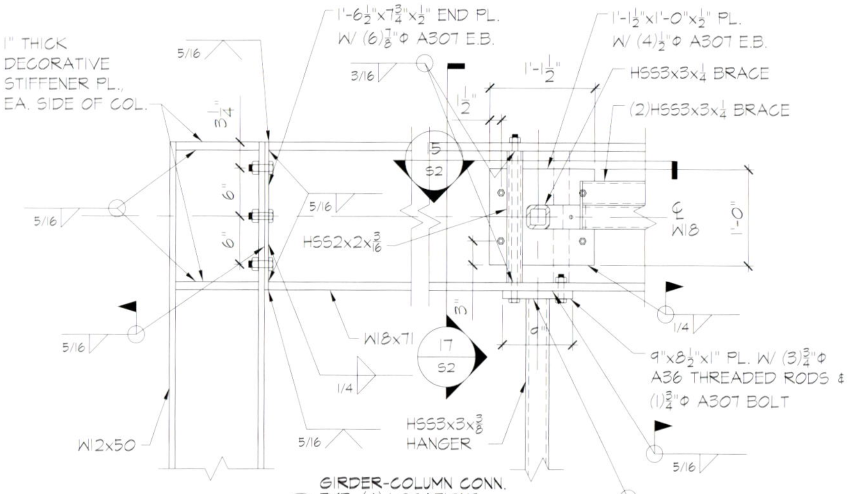


GIRDER-COLUMN CONN. TYP. (4) LOCATIONS
SCALE: 1" = 1'-0"

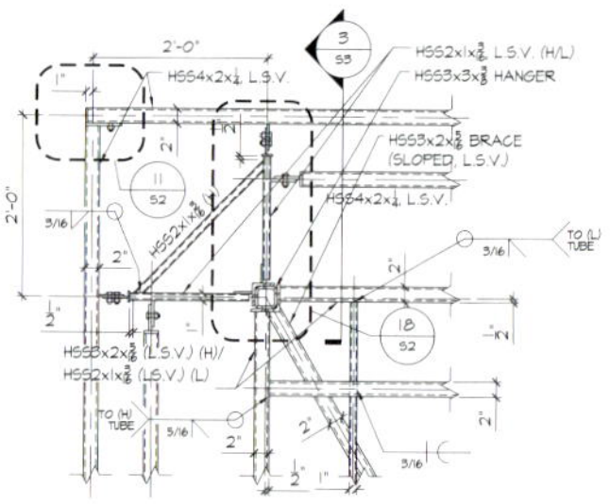


CONNECTION DETAILS
SCALE: 1" = 1'-0"

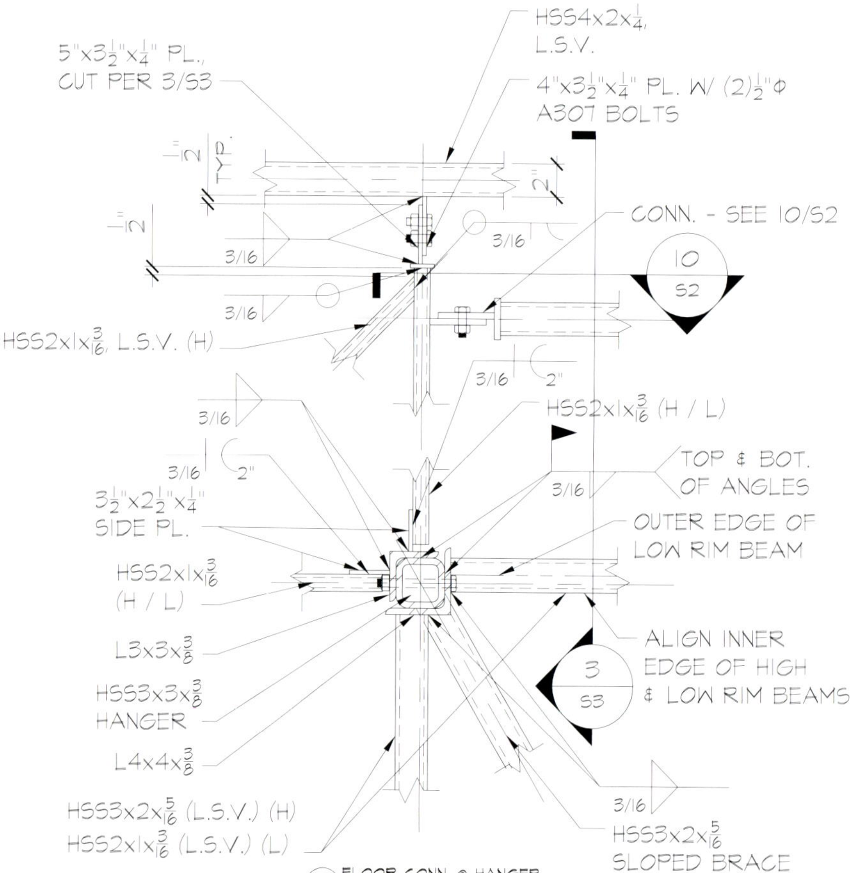


FLOOR CONN. @ HANGER
SCALE: 1 1/2" = 1'-0"

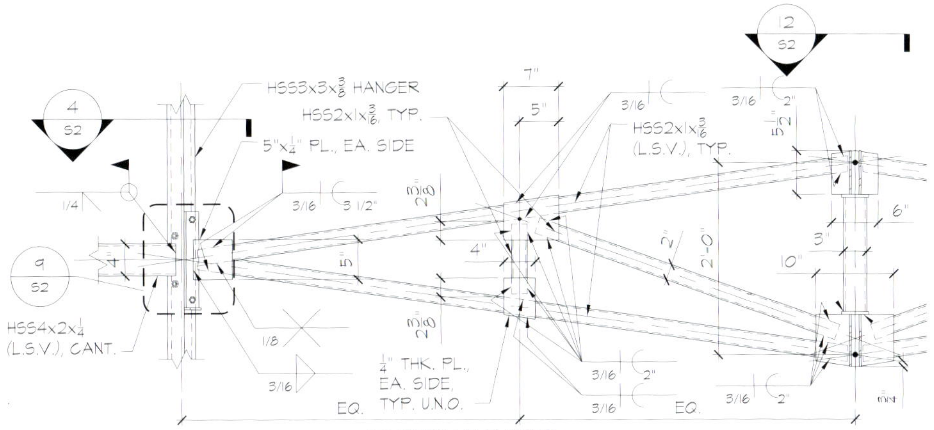


TRUSS 1 ELEVATION
SCALE: 1" = 1'-0"

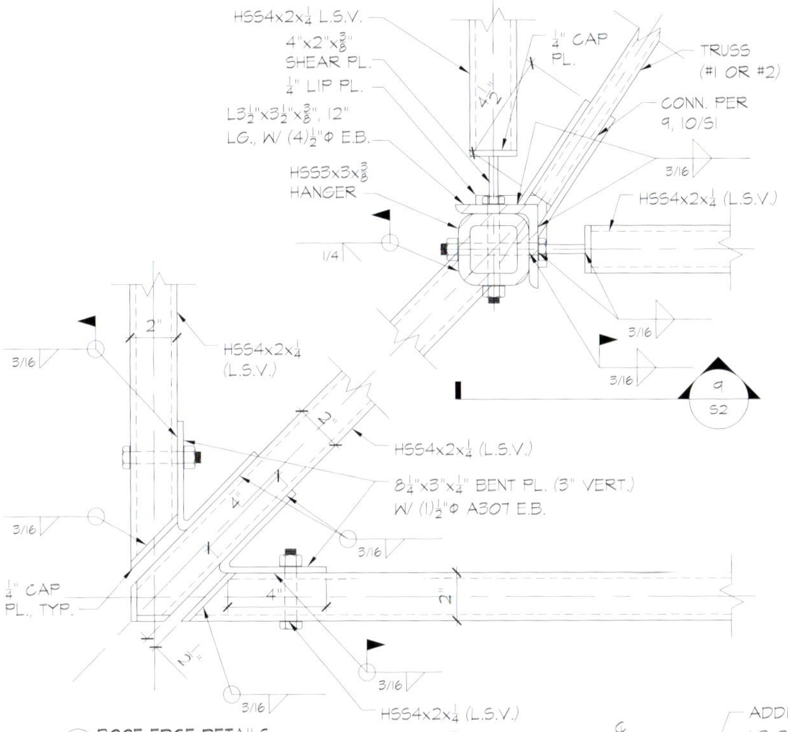


ROOF EDGE DETAILS
SCALE: 3" = 1'-0"

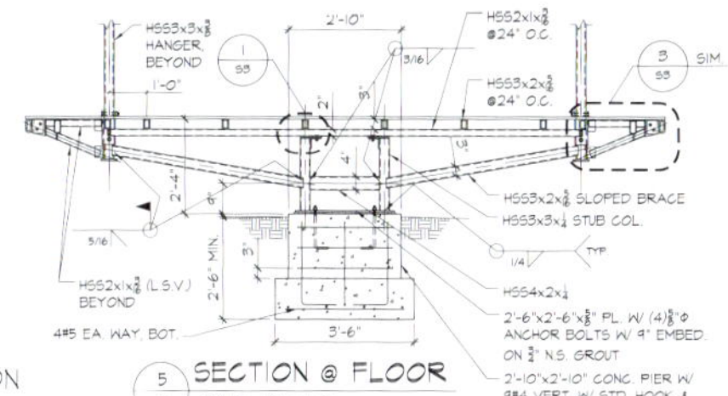

SECTION @ FLOOR
SCALE: 1/2" = 1'-0"

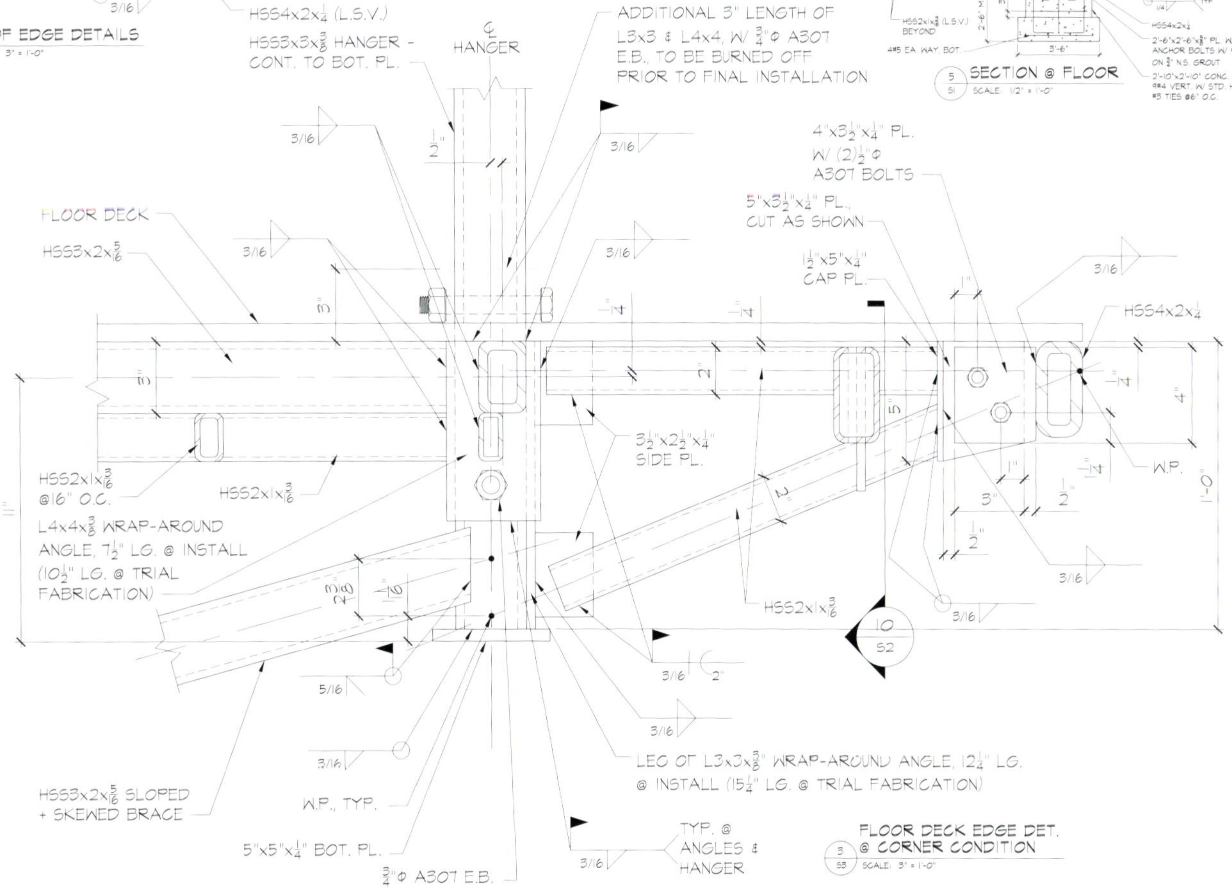


FLOOR DECK EDGE DET. @ CORNER CONDITION
SCALE: 3" = 1'-0"

# 帽子茶室

A1 Architects

“帽子”是我们完成的第三个茶室。这个小房子为主人提供了一个和客人们聚会的空间，也是迷人小花园里一个开敞屋檐下的小避风港。在狭小的茶室内营造出一个舒适的私密空间，同时还要使它具有独特的氛围，这对我们来说是一个很大的挑战。茶室内部空间给人一种与外部世界截然不同的感觉。

“帽子”是我们所设计的三个茶室中最小的一个。高高的屋顶看起来仿佛是小花园的大帽子，茶室也因此而得名。它内部空间的面积为1.8m×1.8m，可以容纳一位主人和两位客人。坐椅是为茶室主人预留的，客人们则可以舒服地坐在由落叶松制成的宽凳上。

帽子茶室的屋顶衬层使屋顶显得很矮，近乎方形的格局向上逐渐演变成圆形天窗的形状。由于空间有限，炉子被设计成一个隐藏在宽凳下方的抽屉。从茶室向外望去是风景如画的花园，而花园正对着斯特拉瓦郊区的一个绿草丛生的废煤堆。移动推拉窗和外面的百叶窗，人们可以看到不同的景色。

材料的选取在茶室的设计过程中起着关键的作用。我们要寻找那些自然老化的材料，这样可以使空间内部保持温暖舒适。经过木工处理的橡木结构竖立在深灰色的石头上，与深色落叶松表面和木瓦屋顶相呼应。内部主要墙体都覆盖了橡木胶合板。整个结构由雕塑家Vojtěch Bilišič在2010年夏天打造完成。

与日本建筑师藤森照信的会面给我们带来了重要的灵感，也影响了我们所有的茶室设计。以前我们在设计茶室时，通常会按照自己的理解来阐释主题，刻意追求与日本传统的不同之处。然而现在，我们会采用传统的材料和中欧的表现形式，把茶室融入到当地的环境中。

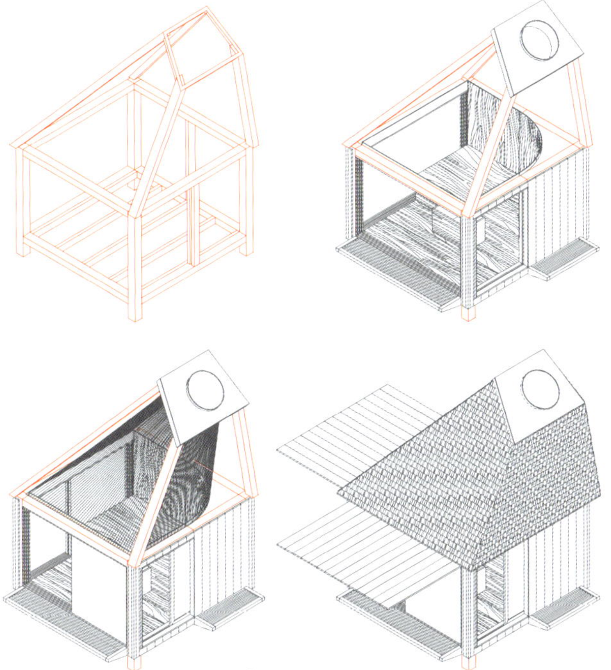

方案设计　schematic design

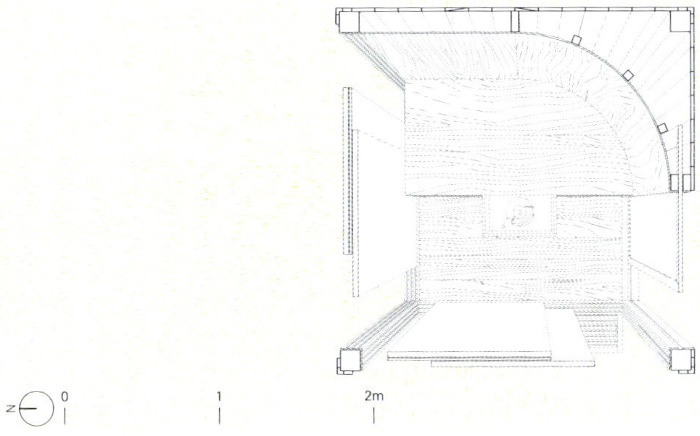

## Hat Teahouse

"Hat" is the third teahouse we have realized. It is a small house, which serves as a place for gathering with guests, and a tiny haven under the open roof in a little charming garden. The minimal space of a teahouse is a great challenge for us to search for intimate and pleasant space, which brings unique atmosphere with its own secret inside. The inner space of the teahouse gives distinct perspective of the outside world.

"Hat" is the smallest of the three teahouses we have already designed. The name derives from its tall roof resembling a big hat in the garden. Its inner space is meant for 2 guests and a host within dimension of 1.8x1.8m. The seating is suit to its owners, so guests can enjoy sitting on the wide larch bench.

The Hat has expressively dwarfed roof lining, the almost square layout transforms upwards into the round shape of skylight. Hearth is designed due to the minimal size of space as a drawer hidden under the bench. The main view is oriented to unique picturesque garden towards bizarre horizon with grassy coal waste dumps of Ostrava outskirts. The view could be adjusted by sliding windows and outer shutters.

Materials play a key role in the design of the teahouse. We search for natural aging materials which make the space warm and livable. Carpented oak structure stands on dark gray stones, which harmonize with dark larch facing and shingle roof. The main inner wall is clad with oak plywood. All the structure was built in summer 2010 by sculptor Vojtěch Bilišič.

Meeting Japanese architect Terunobu Fujimori was an important inspiration for us and it influenced all of our teahouses. In all our previous teahouses, we tried to find our own interpretation of the theme, which intentionally differs from Japanese tradition a lot. We transfer it into a local situation using traditional materials and forms of Middle Europe. A1 Architects

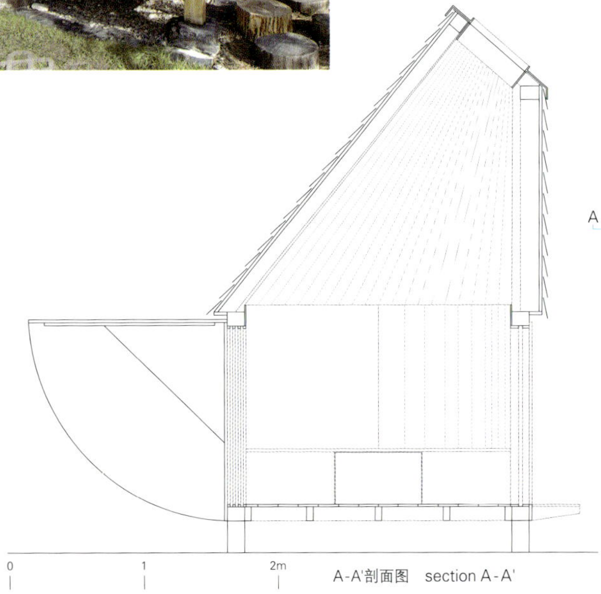


A-A'剖面图 section A-A'

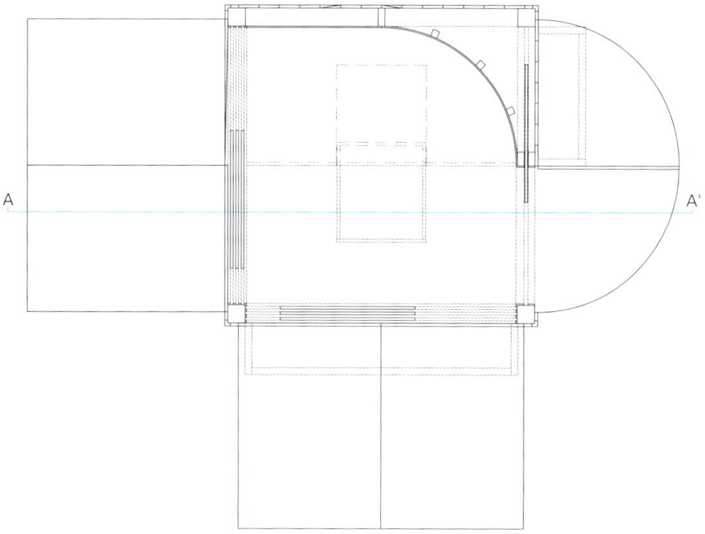


一层 first floor

项目名称：Hat Teahouse
建筑师：Lenka Křemenová, David Maštálka
地点：Ostrava, Czech Republic
木工：Vojtěch Bilišič
建筑面积：3.25m²
竣工时间：2010

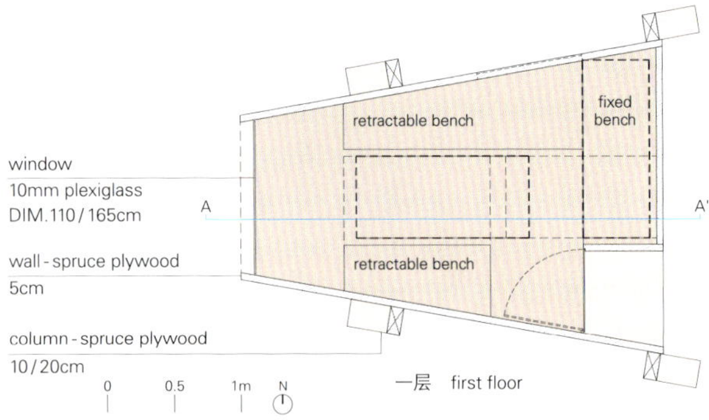


一层 first floor

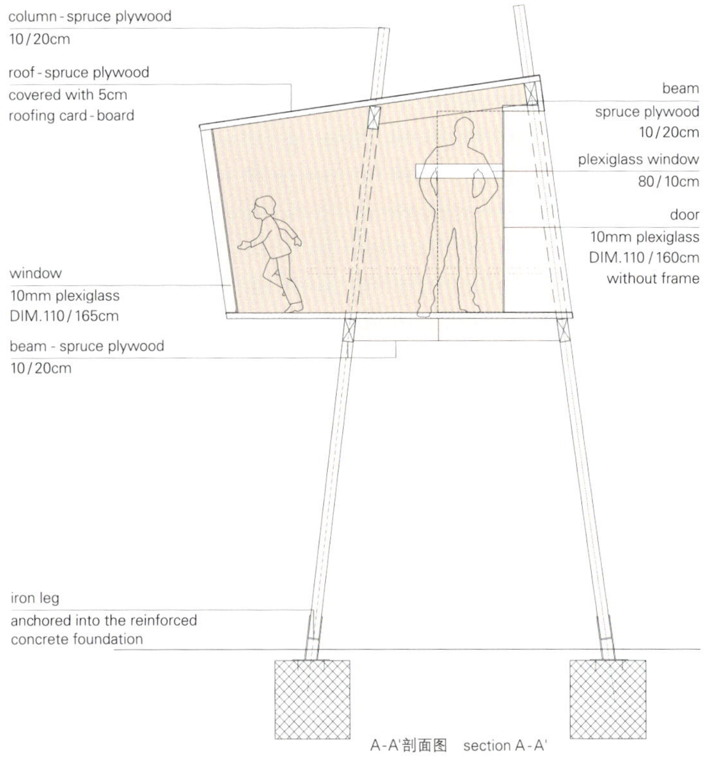


A-A'剖面图 section A-A'

house is something special. It stands on an independent structure and does not bear any load on to the tree's trunk and crown. As such, it carries a deeper message: enjoy nature, be part of it, but leave it untouched.

Its charm lies in the pristine contact with nature and in its primeval dwelling, although it can double up as a watch tower, a hiding place or a spaceship. In it, you can be hidden and invisible or bold and invincible.

The house doesn't seek to emulate classical tree houses which are based on imitations of real houses or miniature garden sheds, but is rather designed using modern principles of design and applying simple, clean lines. Children are therefore brought closer to a different appreciation of shapes and space, which ultimately arouses a relationship with the environment, as the house does not defy nature nor change it, while only standing by and observing it.

The sustainable concept extends to the use of materials. The house is made of glued pine wood, which is protected on the outside with a colorless nano coating. This is supposed to contribute to the wood's longevity, while allowing it to take on its own gray patina. The structure is also an experiment to test the protection of the wooden structure using a nano coating. The roof and the projecting side are covered in cardboard, which successfully defies atmospheric influences. The interior is minimal and thus allows endless changes to scenarios, sceneries and offers limitless new experiences. Nada Serbelj

# 避难所

Petra Gipp Arkitektur

Kivik艺术中心

一座浅白色混凝土结构被镶嵌在树林繁茂的斜坡上。它是如此隐蔽以至于游客很难马上发现它。

整个结构集中在由大片森林围起来的一片空地上；在树林与混凝土之间，一段通道和一个内部空间旋转而生。楼梯切开并穿过整个空间，使整个结构更紧密地结合在一起，同时还与地面相连。通道向上延伸至内部空间，空间中的木材核心筒形成一个长凳似的壁龛。天花板从墙体处向内缩进一点，露出一小片蓝天，光线从切口处洒落下来，从切口进入的雨水和秋叶也成为空间的一部分。室外与室内的界线非常模糊。木材和混凝土材料会随着时间的推移而逐渐老化，所以希望随着时间的流逝这种空间的体验感也会加强。

基础位于建筑后部，这样可以抬升整个结构，不与地面接触。无论在理念上还是在视觉上，设计概念都把建筑对环境造成的影响降到了最低。

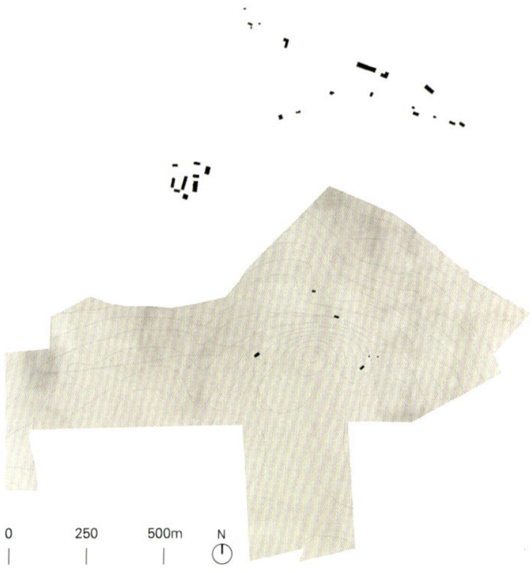


## Refugium

Kivik Art Center

A pale concrete structure is embedded in the lushness of a slope. Secretive, it does not immediately reveal itself to its visitors.

The entire form is gathered around a nave made out of massive wood; between wood and concrete, a passage and an inner space are spun. As a staircase carves its way through the volume, it ties the structure tighter together still, while it connects to the ground. The passage opens up into the inner space, where the wooden core forms a niche as a bench. The ceiling releases from the wall to reveal a thin slice of sky, a cut through which light trickles down and both rain and autumn leaves are allowed to be a part of the space. The boundary between what is inside and what is outside is blurred. The materials, wood and concrete, are allowed to age in the hope that over time, the experience of both space and place will further be enhanced.

Foundation is laid in the back, which allows the structure to rise above the ground and leave a large part of the site untouched. The idea is to inflict as little damage as possible on the site, conceptually as well as visually. Petra Gipp Arkitektur

A-A'剖面图 section A-A'

项目名称：Kivik Art Center
建筑师：Petra Gipp Arkitektur AB
合作者：Petra Gipp, Maria Videgård and Kalle Hjalmarsson
结构工程师：Bengt Nilsson, BN Konsult Lift & Transport AB
承包商：Bengt Nilsson, BN Konsult Lift & Transport AB
甲方：Kivik Art Center, Sune Nordgren
地点：Kivik, Sweden
基地面积：24.6m²
竣工时间：2010
摄影师：courtesy of the architect (©Gerry Johansson)

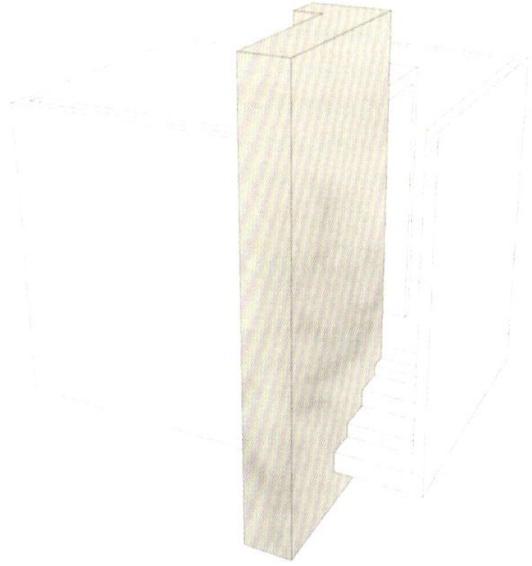

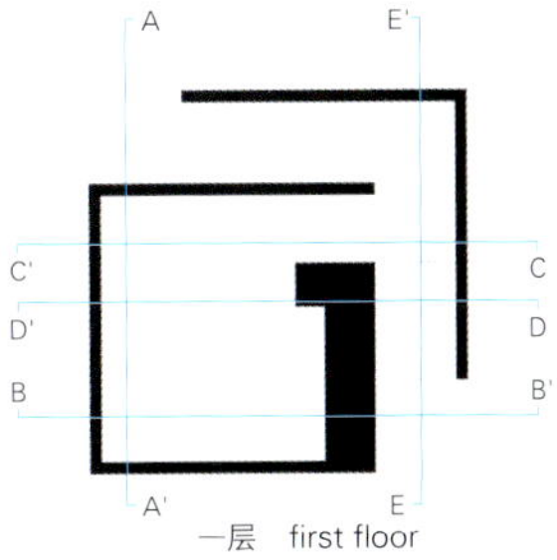

一层　first floor

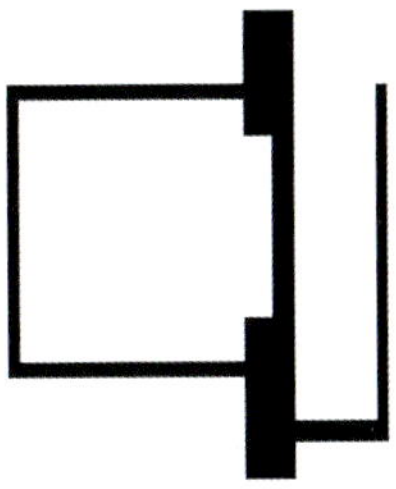
B-B'剖面图　section B-B'

C-C'剖面图　section C-C'

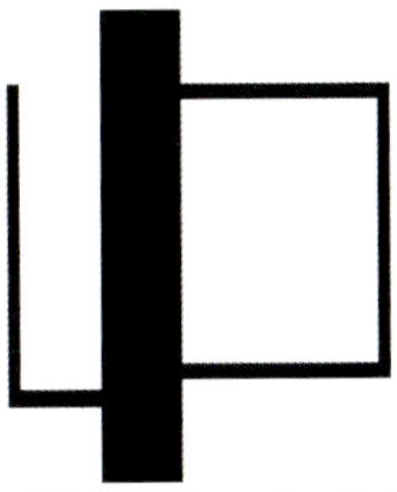
D-D'剖面图　section D-D'

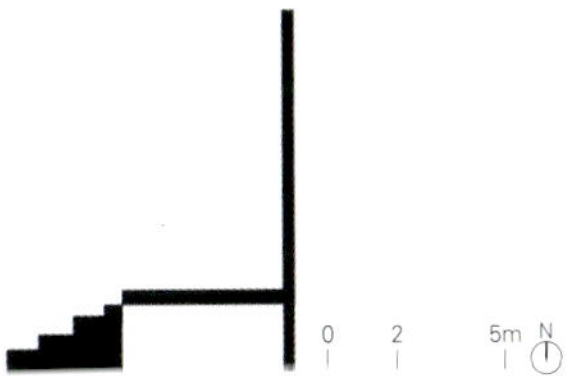

E-E'剖面图　section E-E'

# 木柴堆

Noa Biran + Roy Talmon

“木柴堆”是2011年“温暖的小屋——冰上艺术建筑竞赛”的获奖作品之一，于2011年1月至3月建造，地点位于加拿大马尼托巴省温尼伯市，沿结冰的阿西尼波河的冰道而建。

这次竞赛于2010年10月举办，目的是要求艺术家、建筑师和设计师相互合作，为那些在冰道上滑冰、滑雪和散步的人提供一个临时的休息场所。

生火是取暖的最基本方法。“木柴堆”被设计成一个室内空间，由四面墙体框架所限定，框架里面可以堆放木柴。在小屋中间有一个火坑，滑冰者可以生火取暖。木柴是小屋的建筑材料，“木柴堆”既是燃料的容器，也是取暖的地方。

整个滑冰季节，人们不断地使用木柴，同时也不断地改变着小屋的外貌。进入冬季，人们堆放木柴，木柴堆的高度达到最大，将小屋与外界的冰天雪地隔开。通过木柴之间的缝隙，人们可以看到里面的篝火。春天来临时，木柴堆的高度越来越低，小屋的内部空间也逐渐呈现并曝露在外面。到了夏天，这个无任何覆盖的小屋也可以成为人们在河边躲避阳光的凉亭。

当我们为了建成这个小屋而开始细节设计时，不得不做出了一些调整。因为小屋直接坐落在冰冻的河道上，那么当河上的冰开始融化时，我们需要一个把小屋移走并存放的方案。因此小屋被设计成由多个独立的结构构件组成，其中包括四面墙体框架、地板和屋顶组件。现场安装时，所有的组件只需要用简单的U形螺栓相连接就可以成为一个整体。此外，矩形截面立柱作为中心结构，承担屋顶重量，并通过地板把重量转移到冰面上，同时还可以作为屋顶的排水系统。

当谈论到建筑时，“性能”这个术语通常会涉及结构的功能和操作。但是，性能也可以指建筑表现、呈现的理念。在设计中，我们希望表达两种概念：小屋的外貌会随着时间的推移而不断变化，它在冬天是一个封闭取暖的空间，而在夏天是一个开放的空间。同时这种变化反映了人们对小屋的使用，因此，滑冰者才是整个设计中最重要的部分。

## Woodpile

The “Woodpile” was one of the winning entries in the 2011 “Warming Huts – an Art and Architecture Competition on Ice”, and was installed between January and March 2011 along the skating trail at the frozen Assiniboine River in Winnipeg, Manitoba, Canada.

The competition, which was held during October 2010, called for collaboration between artists, architects and designers to put forward ideas for temporary shelters that would provide a space for skaters, skiers and strollers to rest along the trail.

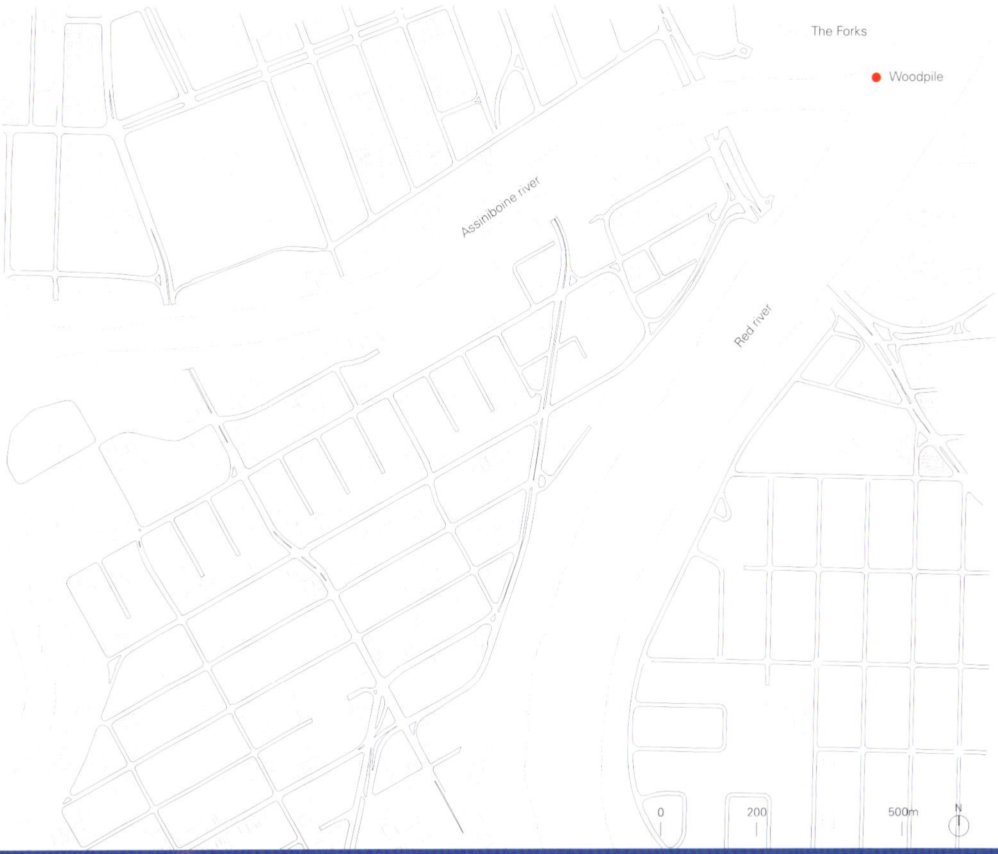
The Forks
Woodpile
Assiniboine river
Red river
0
200
500m
N

项目名称：Woodpile
建筑师：Noa Biran, Roy Talmon
甲方：The Forks Foundation
协调员：Peter Hagraves, Sputnik Architecture
金属加工：Hannes Berger, Alpine iron works
地点：Winnipeg, Manitoba, Canada
建筑面积：12.25m$^2$
设计时间：2010.10～2011.1
施工时间：2011.1
竣工时间：2011.2
获奖情况：winning entry
摄影师：courtesy of the architect (©Brian Gould) - p.75
Courtesy of the architect (©Dave Pancoe) - p.76, p.78
Courtesy of the architect (©El Design Winnipeg) - p.79 top-right
Courtesy of the architect (©Henk Von Pickartz) - p.77, p.79 top-left, bottom

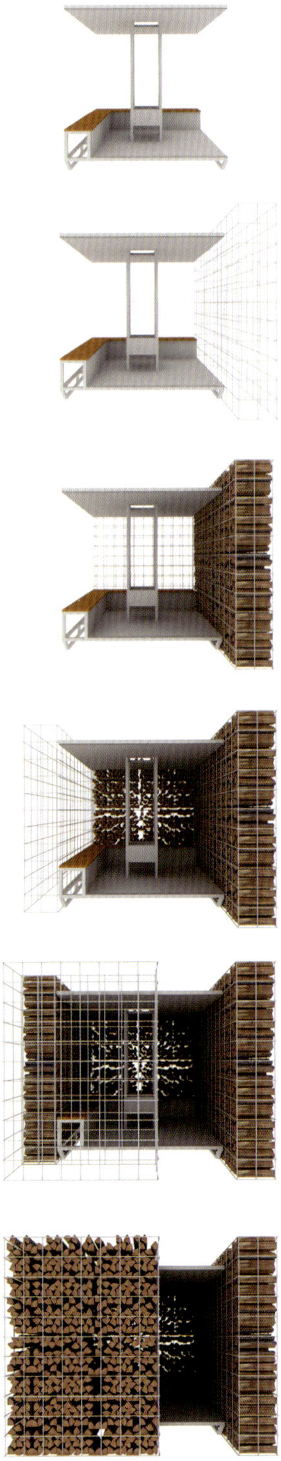

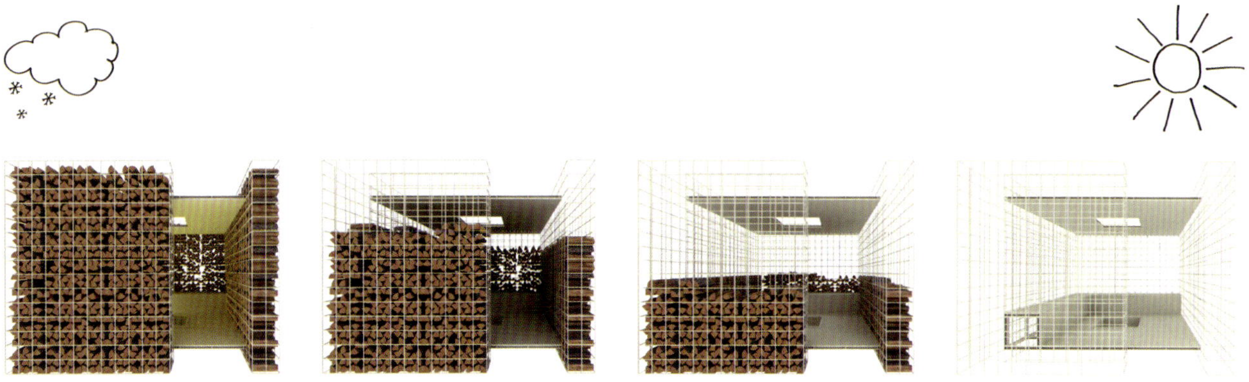

Starting a fire is the most elementary act of warming. "Woodpile" was designed as an inner space, defined by four peripheral wall frames in which firewood is stacked. At the middle of the hut, a fire pit was planned to keep the skaters warm. By using firewood as the construction material of the hut, "Woodpile" serves as a container both for the lighting material as well as for the act of warming.

The use of the firewood through the skating season constantly changes the hut's appearance. As winter begins and firewood is stacked, the woodpile's level is at its maximum, closing the hut from the surrounding and isolating its inner space from the winter colds. Looking through the cracks between the woods, one can see the campfire being held inside. As spring approaches, the stack's level is lower and the hut's interior space is gradually revealed and exposed to the outside. By summertime the hut's naked construction could also serve as a shaded pavilion along the river.

As we approached the detailed planning for the hut's realization, several adjustments were required. Since the hut was to be located directly on the river's ice, a solution was needed for its removal and storage when the river ice melted. The hut was therefore planned with separate structural elements; including four wall frames plus floor and roof components. All elements, when placed together on site, are connected with simple U-bolts and serve as one structural unit. Furthermore, a central construction of RHS columns was designed to carry and transfer the roof's load through the floor elements onto the ice while simultaneously serving as the roof's drainage system.

When speaking about architecture, the term "performance" usually relates to the aspect of function and the operation of a structure. However, performance also refers to ideas of display and presentation. In our design we wanted to express these two notions: As the hut's appearance constantly changes through time, it conceptually turns from a closed and sheltered space during winter to an open structure by summer time. This change reflects the skater's use of the hut so that they play the most significant role in its design. Noa Biran + Roy Talmon

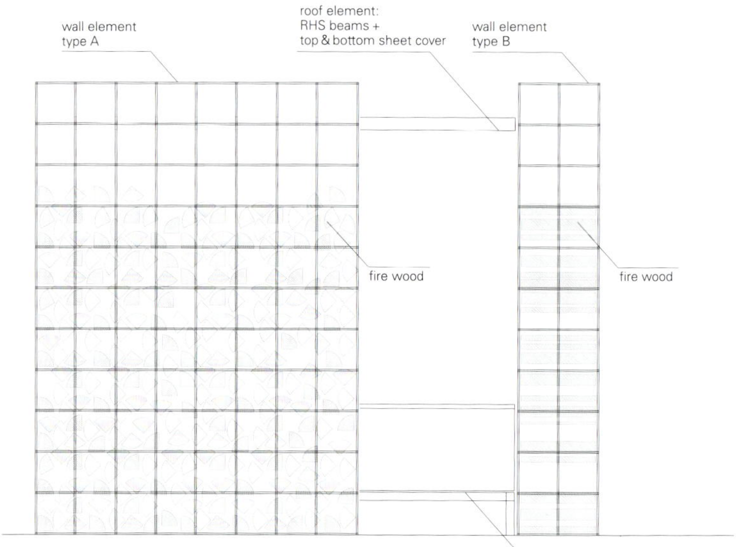


南立面 south elevation

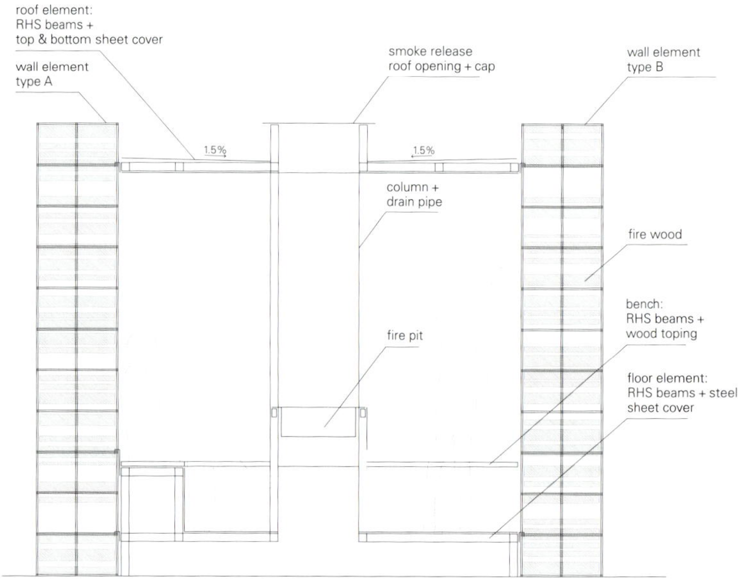


A-A'剖面图 section A-A'

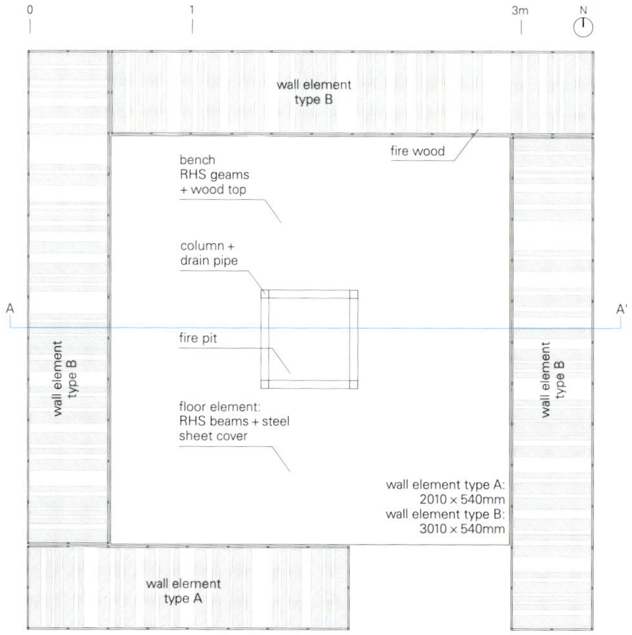


一层 first floor

CCM

# 十字门

Ivo Pavlik

十字门所在的位置原来是一个岔道口，位于一座山丘上。十字门的名字叫"U kamennkéo kříže"（在石头十字架处）。自1818年至2005年，这里一直坐落着一个石头十字架。该石头十字架在城镇的周围处于最高点上。由于频繁犁耕土地，旧十字架根基不稳，被搬到了教堂附近的镇上。U kamennkéo kříže的基地之前是一个岔道口，三条路在此交汇——一条通向Dukovany镇，还有一条通向Hermanice村。这些路在修建核电站的时候被损毁了。设计的初衷是将十字架放回原来的景观中，安置在它之前一直所在的位置上。

十字架是十字门的一部分，除了其自身原有意义外还有另一个新的象征意义——生与死的假想界线。门后向东、向西皆可行，跨越这道门意味着"到另一边去"，到能使我们冥想的地方去。这道门通过一条古道与一个一直存在于此的公墓相连。

设计灵感源自人们习惯于将骨灰撒在离人近的风景优美的地方。十字门的雏形源于3D草图，但是最终成形却是源自历史和地域特征。十字门的墙和过梁由混凝土浇筑而成，外表面由稻草构成，这些稻草在2010年1月10日下午一点被烧毁。因为大火的缘故，新的结构和墙体颜色变成了现在的样子，而且稻草上紧粘着灰烬。墙的东侧与过梁形成了一个十字架，表面裹上金箔，十字门至此完成。

过梁的位置稍作改变，就形成了一座极具史前感觉的墓石牌坊。这座石十字架附近就有一处史前遗址，人们在那里发现了众多史前人类使用过的工具。

十字架墙体上面部分被切下来一部分，放在了大树的后面，人们可以在此处休息，而且我们还可以在上面放蜡烛或鲜花。如果我们想到切割下来的石墙一侧，就必须经过大树。大树阻挡了人们的视线和脚步，让我们不能直接看到或到达十字门的另一端。只有在门的正前方才能看出十字架的形状来，从后方也隐约可见，只是另一座墙与树融为了一体。坐在树下的"长椅"上时，可以看到南面的开阔景象、北面的小镇以及延伸至脚下的小路。

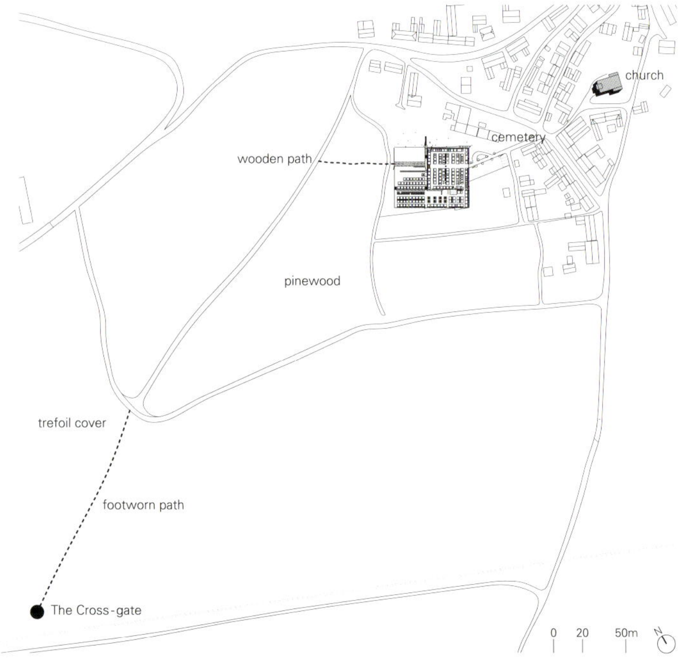


## The Cross-gate

The gate is located at the site of a former junction, on a hill. The name is "U kamennkéo kříže" (At Stone Cross). A stone cross was here since 1818 to 2005. It's the highest point in the town's vicinity. The former cross was moved to the town, near the church, because it became unstable in the frequently plowed soil in the field. The U kamennkéo kříže site used to be a junction. Three paths intersected there – one led from the town of Dukovany, the second one to the village of Hermanice. These routes were destroyed during the construction of a nuclear power plant. The idea is to return a cross back to the landscape, to the place where it had always been.

The cross is part of the gate which carries its own meaning and new symbolic – it's an imaginary border between life and death. It's a place behind which it's possible to get dispersed, in the east-west direction. To walk through the gate means "to get to the other side" where we can contemplate. The gate is related to a cemetery; it's connected to a path from the cemetery which has always existed there.
The project comes out from the original idea of dispersion ash of human being to the landscape on a place that is close to human being.
The base shape of the first idea is recognized from 3D sketch but final shape of gate in Dukovany comes out from history and place.
The walls and the lintel were casted of concrete into boarding of straw bales, which were burned at 1pm, on the 10th of January, 2010. New structure and color of the walls came to existence, because of fire. Also bales hanged to ash. Eastern side of the wall, which is creating cross with the lintel, was gilded by pieces of laminar gold. The cross is done by this event.
The lintel creates with its location an idealized prehistoric dolmen. There was a settlement in Paleolithic nearby, and many tools of prehistoric people were found there.
A part of the cross' upper wall, which is behind the tree, is cut out and enables sitting under the tree and we can place a candle or flowers to it. If we want to come to it, we must pass the tree. The tree makes it impossible to see through the gate and to move directly. The cross is readable only from the front side. It's also visible from the rear where the other wall is aligned with the tree. While sitting on the "bench" under the tree, our view is southward into the open landscape or northward toward the town and the path which took us here. Ivo Pavlik

项目名称：The Cross-gate
建筑师：Ivo Pavlik
项目团队：Lucie Chytilova, Dana Novakova, Ilona Zitnakova, Ondrej Vosatka
地点：Dukovany, Czech Republic
基地面积：4m²
建筑面积：40m²
设计时间：2007.12
竣工时间：2010.1

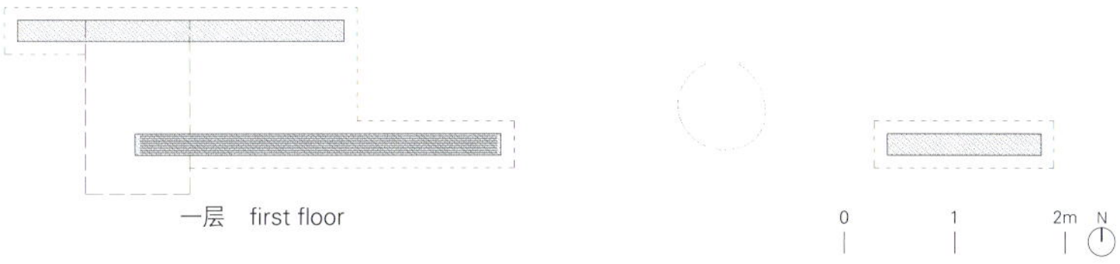


一层　first floor

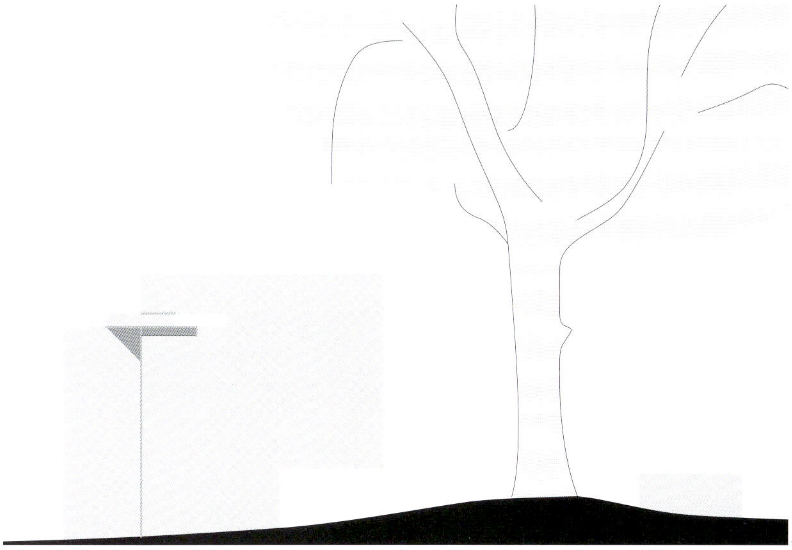

南立面　south elevation

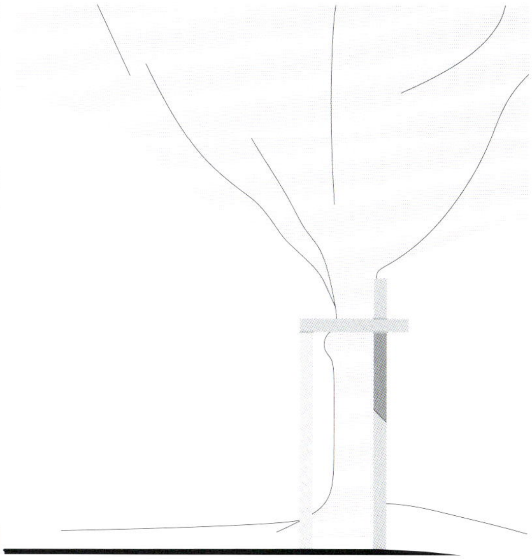

west elevation

# 住宅和城市景观
# Housing and Urban Landscape

直到今天，社会住宅仍是建筑领域中一项复杂的研究课题，尤其是现在的建筑师都在设法将这一学科作为一种工具来研究现代社会的变化。在最先进的案例和项目研究中，最有趣的现象是建筑师们就一种实用程序的组合所展开的诠释，旨在强调事物、城市和景观规模之间相互作用的过程。

实际上，在这里所提到的七个项目的分析中，我们注意到一个组成部分（与规模有关）的存在就能将全部转变为景观项目：所有这些系统能够使建筑超越其本身的物质形态，从而使建筑成为一种改变城市景观的工具。

我们也可以观察一下两种研究方向：对类型学的深刻反思和当代城市大厦的某种变化。由C. F. Møller Architects、O' Donnell & Tuomey和Caramel Architekten完成的项目深刻地说明了类型学概念中的变化以及集体住宅的构成。与此同时，在Garduño Arquitectos、Magén Arquitectos、ZigZag Arquitectura和Frédéric Schlachet Architecte的作品中也强调了住宅和上空空间的关系，因为它们在城市的公共与半公共空间的使用中创造了新的等级。

通过这种方式，新形式的住宅和社会住宅成为理解现代城市发展的最重要的参数。

Social housing is a complex field of research in architecture even today, especially given the use to which architects are now seeking to put the discipline as a tool in investigating changes in contemporary society. In the most advanced case studies and projects the most interesting phenomenon seems to be the evolving interpretation by architects of the combination of a functional program with an intention to emphasize the processes marking the interaction between the object, the city, and landscape scales.

In fact, in analyzing the 7 projects here proposed, we note the presence of a component (related to the scale) that converts all into landscape projects: They are all systems able to take architecture beyond the status of mere object, making it a tool for changing the urban landscape.

We may also observe two investigation lines: the profound rethinking of typologies, and certain changes in the contemporary configuration of the urban block. The projects by C. F. Møller Architects, O'Donnell&Tuomey and Caramel Architekten illustrate deep changes in the concepts of typology and the composition of collective housing. At the same time the works of Garduño Arquitectos, Magén Arquitectos, ZigZag Arquitectura and Frédéric Schlachet Architecte emphasize relationships with the void as they create new hierarchies in the use of the city's public and semi-public spaces.

In this way the new forms of dwellings and social housing are among the most important parameters in understanding how the contemporary city works.

## 内部景观

集体住宅的设计是否还有实验的空间？如果我们回顾过去几年所发生的事情，就可以得到以下结论：尽管房地产市场存在着压力，人们不断地思索着社会的发展与变化，但住宅这一主题一直是现代设计研究中至关重要的因素。

在集体住宅这个问题上，我们不仅谈论其构造和创造形式的演变过程，同时也关注当代设计师是如何解释功能性项目的变化，以及住宅与当代城市外形的关系。

在建筑形式方面，我们可以看到项目是如何日益发展成为工序流程或策略体系的产物的，这种体系以一种新型的、意想不到的方式来定义形式结构，而不只是在构造上理解设计概念，将其作为整体与部分之间的平衡。

纵观当代设计理念中最具标志性的作品，我们发现这些项目实际上在建筑、城市以及景观的关系系统中更具有识别性。这种现象已在近年来的作品中显现出来，如MVRDV为老年人设计的老年公寓、由荷兰建筑师Willem Jan Neutelings和Michiel Riedijk联合设计的锡塔德市的商住综合楼、阿姆期特丹Java Eliand桥墩边的LJ高楼、由Kazuyo Sejima设计的位于Gifu Kitagata的公寓楼、近年来由Bjarke Ingels（BIG）与JDS设计的建筑大楼“山阶住宅”以及后来由BIG单独设计的“8字之家”。这些案例都进一步说明了这个主题，按逻辑学能够跨越类别等级，用已建立的硬性规定去维护这种学科的常规惯例。这种方式在建造复杂的、多层结构时没有正式的文书，比如图表。这些建筑本身也收获了来自城市共有空间的特点，同时日益个性化的居民需要共享他们的生活空间，通过解释这种观念来显示公共与私人空间的关系。

社会变革使功能性项目得以改变，使现代的集体住宅远离过去的建筑样式。我们远离了诸如由勒·柯布西耶设计的法国马赛的集体住宅，或者一些结构试验品，像罗宾汉花园、由阿利林和彼得·史密森设计的伦敦波普勒市政综合住宅或由几位意大利建筑师联合Mario Fiorentino共同设计的罗马一公里建筑——Corviale。这些设计摒弃

## Domestic Landscapes

Is there still space for experimentation in the design of collective housing? If we look at what has happened over the last few years we can conclude that, despite the pressures of the real estate market and constant speculations concerning societal progress and change, the topic of housing continues to be crucial in any exploration of the limits of contemporary design.

In the case of collective housing we may speak not only of how composition and invention of form have evolved, but of how contemporary designers have addressed changes in functional programs and the relationships between the dwelling and the shape of the contemporary city.

In relation to the form, we can see how projects are becoming increasingly a product of processes, or the result of strategic systems that define formal structures in new and unexpected ways, rather than settle for understanding ideas on composition simply as a balance between the parts of a finished object.

Looking at some of the works that have most strongly marked the evolution of contemporary design thought, we discover, in fact, that these projects are more identifiable as systems of relations between different scales: the architectural, the urban and the landscape. This phenomenon has been manifest in recent times in such works as the Wozoco's apartments for elderly people by MVRDV or the Prinsenhoek residential complex in the city of Sittard and the IJ Tower on the edge of the pier Java Eliand in Amsterdam both designed by Dutch architects Willem Jan Neutelings and Michiel Riedijk, or as the Gifu Kitagata apartment building in Gifu by Kazuyo Sejima and, more recently, in such buildings designed by Bjarke Ingels(BIG) with JDS as the Mountain, and later in the 8 House, designed by BIG alone. All are examples of an approach to the subject in terms of a logic that allows for jumps in scale involving categories that disciplinary conventions have worked to preserve as established and rigidly defined. It is an approach that acts with no formal instruments, such as diagrams, in building complex and layered structures. These are buildings that receive in themselves the characteristics of an urban space working on a collective space, and that take on the emerging relationship between public and private space by interpreting the idea of an increasingly individualized resident who nonetheless has to share living space.

照片提供：ChangYeol Choi

勒·柯布西耶设计的位于菲尔米尼的集体住宅，1965年

了过去几年出现的错误观念，带给我们大量的建筑构造，其社会结构和联系严格按照建筑师的意愿去建造，现代集体住宅像景观建筑一样层出不穷，它并没有严格界定的等级。

集体住宅已成为开放的系统，有着更加灵活的内部空间，就如同将不同的子类型聚集在一起，而这些子类型是依据对空间的使用性、技巧性和特征的整合建立起来的。因此集体住宅成为一种能转变城市范围的体系，今天有着越来越多的复杂性和不确定性。例如，它不再视城市与乡村为截然相反的方案，从而取代了景观，使不同的项目可以求同存异：历史的结构、郊区、大型景点和基础网络设施创建了时而自治、时而重叠、时而中立的组织结构。这种混搭的方式早已将城市变成一个缺少功能的景点，根据既定建筑的逻辑连续性，变得很难控制。如雷姆·库哈斯在他的著作"重大和普通的城市"中所推论的：在这种情况下，开放的空间失去了公共空间的内涵，多变且非连续性的实体产生的功能使城市组织逐渐消失掉，就相似的逻辑性或前后的连续性而言变得越来越自治、中立。

集体住宅的改变不仅涉及形式和类型的发展，而且就城市大厦提出了一项新的设计方案，使其成为一个更有渗透性的整体，室内空间不但影响了在拥有光和空气的条件下的实用逻辑性，而且对缺少城市公共空间这一问题采取了弥补措施。新型的混合空间的建造（半私密？半公开？）要求这些空间发挥出作用，否则就在城市中消失。这一过程有选择地进行着，社交场所从来没有完全开放；它们被密切关注并保护，它们与城市的关系有着不完全性、自发性和临时性的特点。

因此，这里分析的七个项目清楚地说明了两种趋势：对城市大厦的一种现代定位和对住宅类型学的深刻反思（这里包含了社会住宅）。第一种趋势强调大小、城市空闲的公共空间与半公共空间的关系，其代表有Garduño Arquitectos、Magén Arquitectos、ZigZag Arquitectura和Frédéric Schlachet Architecte的作品。O' Donnell + Tuomey和ZigZag Arquitectura的项目也综合地展现了两种类型——形式的重新设计和对城市大厦的重新定义。在第二种趋势中，我们可以看到集体住宅的类型学观念中的变化和转变，此种变化和转变体现在C. F. Møller Architects、O' Donnell + Tuomey和Caramel Architekten作品中。

Social change thus mutates the functional program, moving contemporary collective residences away from the examples of architecture that history has left us. We are far from the Unité d'Habitation in Marseille by Le Corbusier or megastructural experiments such as Robin Hood Gardens, the council housing complex in Poplar, London, by Alison and Peter Smithson or the Corviale, the one kilometer Building in Rome, designed by a group of Italian architects coordinated by Mario Fiorentino. Detached from the mistakes made in those years that brought us very large architectural structures in which social structures and relationships were rigidly structured according to the will of the architect, contemporary collective dwellings appear more and more as building-landscapes in which there are no rigid, defined hierarchies.

They have become open systems with a greater versatility of interior spaces, as well as a cohabitation of different subtypes organized according to combinations that articulate the uses, techniques and features of spaces. Collective housing thus becomes a system able to transform an urban territory, today increasingly more indefinite and more complex, such that it can no longer be recognized as a scenario in which the city and the countryside are distinct and opposing categories, having become instead a landscape in which different orders coexist and may be compared: the historical fabric, the suburbs, large attractions, and infrastructural networks create textures at times autonomous, sometimes overlapping, occasionally indifferent. This intermingling has long been turning the city into a weak landscape that is difficult to control according to the logics of continuity in a given building. It is a situation in which the open space loses its connotations of public space as the urban organization is dissolved into areas produced by heterogeneous and discontinuous entities increasingly more autonomous and indifferent to the logic of proximity or contextual continuity, as theorized by Rem Koolhaas in his writings on Bigness and the Generic City.

Mutations in the collective housing project involve not only formal and typological development but propose a new design of the city block that makes it a more permeable entity, in which the interior space not only responds to the functional logic of providing light and air, but assumes a status that compensates for the lack of urban public space. Construction of these new categories of hybrid spaces (Semi-private? Semi-public?) stakes a claim for these spaces as playing a role otherwise absent in the city. This process is taking place selectively. The places for sociability are never completely open; they are monitored and protected by buildings that are granted only partially, voluntarily, and provisionally to a relationship with the city.

Therefore, the seven projects here analyzed define quite clearly

照片提供：Rory Hyde

MVRDV设计的位于阿姆斯特丹的老年公寓，1997年

照片提供：Lemondroid

位于罗马的Corviale住宅，11层，是一座1千米长的单独建筑，可容纳1200间公寓套房和6000人，1982年

根据这些分类，社会住宅的新形式是理解当代城市结构的最重要的参数。

**第一部分：城市建筑的新策略**

44户社会住宅——Magén Arquitectos

Magén Arquitectos的项目以一座黑白两色的大型建筑的形式展现在人们面前，用黑色条纹标志着周围边界，形成一个清晰的U形。乍一看这个设计方案似乎是对一个典型的结构予以了特权，目的是将建筑建在地区边缘处，以此创造出一个真空的私密空间。然而，正如建筑师所指出的，项目的宗旨是要在城市、公共空间和私密（或半公共）空间之间建立一种联结方式。在此背景下，最有趣的是项目的规模。除建筑的形式方面，建筑师还对建筑和城市的构成予以考虑并加以平衡，这种真空的解读是最有吸引力的。这里的上空空间实际上是一个转变的场所，联结城市结构的同时也联结了建筑内的私密空间。为达到这种效果，建筑师们将标志庭院入口处的结构悬挑起来，这种方式突出了视觉效果，以创建一个可直接察觉到的入口通道，这样在建筑之内就可以看到外部空间。因此，在庭院中两种不同模式的使用得以重叠：外面的人们可以从此进入建筑，这个空间还被组成集体住宅的各个单元当成一个普通的庭院来使用。所以它成为城市中的一块聚集地，这个小正方形的轮廓很清晰，城市规模在其中显得很小。与此同时，它起到了一定的联系作用，能够使住宅空间向外延伸，给居民的社会生活带来便利。建筑沿着基地外围的连续性增加了人们对城市中这一固定位置的感知度，从而促进了庭院作为公共空间的使用，这种公共空间的规模在建筑之外会大大消失。

米耶雷斯社会住宅——ZigZag Arquitectura

ZigZag Arquitectura的住宅楼也建在基地的外围，创建出一个大型的内部空间，其传统的模式在外形上与Magén Arquitectos的运用手法有相似之处。然而，这个设计理念源于地点与项目之间的解读问题，这种解读更为复杂。通过图表仪器的使用，ZigZag Arquitectura采用了一种方法，将城市规模的住宅与周围景观联系起来。这种方法通过不同的层次来展现：首先，设计出综合体，使立面表

two trends: a contemporary definition of the urban block and a profound rethinking of the residential typology (in which we include social housing). Representing the first group, which highlights the relationship between size and the empty public and semipublic spaces of the city, belong the projects of Garduño Arquitectos, Magén Arquitectos, ZigZag Arquitectura and Frédéric Schlachet Architecte. The projects of O'Donnell + Tuomey and ZigZag Arquitectura are also an expression of a synthesis of both categories, in that they work on both redesigning form and redefining the urban block. To the second group, in which we can read the changes and transformations of the typological concept of collective housing, belong the projects by C. F. Møller Architects, O'Donnell+Tuomey, and Caramel Architekten.
Based on these classifications, the new forms of social housing are among the most important parameters in understanding the structure of the contemporary city.

**Part one: New Strategies for Urban Blocks**

44 Social Housing_Magén Arquitectos

The project of Magén Arquitectos appears as a large block in black and white, with black stripes marking the perimeter with a clear U shape. At first sight the design solution seems to privilege a classic typological scheme aimed to set the building on the edge of the site so as to create a vacuum within that becomes a private space. However, as pointed out by these architects, the project's intention is to create connections between the urban, public space and the private (or perhaps semipublic) space. In this context, the most interesting dimension of the project emerges. Beyond the formal aspects of the object, the architectural and urban composition of which is considered and balanced, it is the vacuum reading that is most fascinating. This void is in fact a place of transition and connection between parts of the urban fabric and, at the same time, a private space within the building. To achieve this effect, the architects have suspended the volume that marks the entrance to the courtyard, in this way highlighting the visual perspective so as to create an immediately perceptible access that draws the outside space inside the block. Thus, in the courtyard two different levels of use overlap: the external flow of people moving into the city, and the use of that space as a common court for the units that constitute the block of collective dwellings. It is therefore a meeting place within the city, a small square whose contours are visible, and through which the urban scale is rendered smaller. It is at the same time a site of relationship, which allows the residential space to expand outside and facilitate a social life for the inhabitants. The continuity of the block on the remaining parts of the perimeter enhances a perception of a secure location within the city that promotes

44户社会住宅的过渡性公共广场，Zaragoza

层产生不同的感知度。构成外墙与屋顶特征的表层是由灰色镀锌钢板制成的，可直接与城市联系起来；另一方面，内墙用延伸的木质遮阳板设计而成，它可以根据居民的需要随时被打开或关上。

立面的设计采用了不同的材料，产生了不同的感知度：外面结构看起来是坚硬的、不可改变的，而里面结构看起来是易改变的、舒服的。所谓外部的规模就是城市建筑，而内部给人的感觉像是瞥见了一个小镇，几乎是一个乡村的规模。外部表层的坚硬度似乎是为了保护内部结构不受城市的干扰和人类的接触。

出于相同的意义，外表层所选的材质也起到了隔声屏障的作用。该地点的周边也不是按照封闭的图形设计的，它是开放的，以方便行人在指定的城市道路上通行；建筑高度的设计是为了方便利用室内的光线将内部庭院在视觉上与围绕城市的群山美景相联系，使其规模得到进一步提升，将建筑与城市外部的景色连接起来。建筑的外形似乎与远处山脉的轮廓相得益彰。当人们解读这一综合体时，该建筑产生了各种各样的韵律。

分层的概念产生了一个非常复杂的系统，它将城市建筑的概念重新变成了一种新形式，检验场地与上空空间的构造，在建筑物与景观之间的规模上实现了连续不断的飞跃。

Monte Elbruz公寓——Garduño Arquitectos

在Garduño Arquitectos的Monte Elbruz公寓案例中，两座建筑的对立面和内部上空空间的关系创造了一个联系空间，在此联系空间中可以看到我们所分析的项目中有一个恒定的量是不变的。Garduño Arquitectos的建筑像Magén Arquitectos和ZigZag Arquitectura的项目一样，扩大了内部上空空间，在这种情况下，以一种私密的方式去发掘相关的部分，似乎仅在建筑或城市嵌入结构内有这种可能，而不会出现在建筑的外部。

但是在Monte Elbruz公寓案例中，尽管庭院被建在两座建筑之间的一个纵深的切口处，但城市具有开放性，所以与街道可以产生直接的视觉联系。建筑内部的上空空间本身已经建成，并被分成几个较小的露台，供一楼公寓使用。这些设施没有任何公共职责，但很明显是供每个住户使用的。大楼的透明立面很有优势，用于排去阻塞物，是

its use as a public space, a dimension which is largely missing outside the perimeter.

**Mieres Social Housing_ZigZag Arquitectura**

This block of residences by ZigZag Arquitectura is also set on the perimeter of the block to create a large interior space in a traditional pattern apparently similar to that used by Magén Arquitectos. The design strategy, however, is based on a more complex reading of the relationship between the site and the program. Through the instrument of the diagram, ZigZag Arquitectura applies an approach that links the residential block on the urban scale with the surrounding landscape. It acts by layers: first, the complex is designed so that the skins of its facades create different perceptual dimensions. The exterior skin characterizing the external walls and roofs is made of gray galvanized steel and relates directly to the city; on the other hand, the internal walls are designed as a continuous wooden brise-soleil, which, as it opens up according to the needs of the inhabitants, changes continuously. The difference in materials used for the design of the facades creates a variety of perceptual dimensions: the outside appears hard and immutable while the interior seems mutable and comfortable. The external dimension is that of the urban block, while the perception from inside the court seems to occasion a glimpse of a small town, tendering an almost rural scale. The hardness of the exterior skin seems to protect the internal dimension, the one in which people come into contact, from the aggression of the city.

In a similar sense, the material chosen for the exterior skin also acts as a sound barrier. The perimeter of the site is not designed as a closed figure; it is open to facilitate the passage of pedestrians in specific urban pathways, and the building heights are conceived to facilitate interior daylight and to link visually the internal courtyard with the mountainous landscape that surrounds the city, occasioning a further leap in scale that ties the building to the landscape outside the city. That profile of the building appears to be built in response to the mountain profiles in the distance; it produces a variety of rhythms as one reads the entire complex.

The stratification of the conceptual layers builds a deeply complex system that reworks the concept of the city block, checking the construction of space and void, and fostering continuous jumps in scale between architecture and landscape.

**Monte Elbruz Building_Garduño Arquitectos**

In the case of the Monte Elbruz Building by Garduño Arquitectos, the relationship between the opposite faces of the two blocks and the inner void creates a space of relationship that we have seen to be a constant in the projects we are analyzing. The building of Garduño Arquitectos, like those of Magén Arquitectos and

米耶雷斯社会住宅的可移动的木质百叶窗，阿斯图里亚斯

动态的、开放的，用以弥补外部关系的缺失。

在这个意义上，迄今为止讨论的这三个项目和接下来要讨论的Frédéric Schlachet Architecte的项目都说明了当代城市现有的建筑结构作为人们之间联系和会面的主要场所，其公共开放空间的概念正在慢慢地消失。但从逻辑学角度考虑，这些作用却没有消失，反而在其他空间中发挥更大的作用，比如购物中心和博物馆。因此，城市建筑的内部庭院的功能不再是单一地确保光线和健康要素，而是已变成了有社会规模的私密空间，或者像前面两个案例一样，在建筑物和城市之间成为一个可行的入口。

这些地方都没有完全公开，通过城市的流动，只有部分交叉重叠，甚至决定它们会允许何种程度的违规行为。当代生活条件极大地影响着集体住宅建筑的观念（包括社会的衰退）。

社会住宅综合体，巴黎——Frédéric Schlachet Architecte

在Frédéric Schlachet Architecte的社会住宅综合体里，我们发现建筑空间与私密庭院之间有着相同的逻辑性和相互作用。在这个法国建筑师事务所的项目中，关于在建筑内插入建筑附属物有一个有趣的发展过程。

该综合体在围绕庭院的体量内形成了一个高效、有活力的建筑，并且面对另一个开放空间。建筑师解释说："该建筑群被分为两个花园：一个服务于地块的整体组织布局，另一个确保建筑的开放性（包括俯视邻近住宅），打破现存其他建筑的山墙模式。"这比前面提到的项目都要繁杂：该体量由预制混凝土楼板建造而成，而且内部冷峻、简约的风格也使整座建筑充满魅力。

Schlachet的项目与这里探讨的其他项目不同，该项目似乎是在调整建筑和成本之间的问题，而且可能还要面临一个更紧张的预算。但这种决策并不影响建筑的审美品质，而实际上采用了一种直截了当的表达方式，在设计庭院时也没有任何让步。不管怎样，在定义开放空间时建筑的品质和清晰度并没有丢失。在这里提到的社会住宅项目中，Schlachet的项目最能突出设计者的能力，它以一种简单有效的方式说明了建筑品质不一定与其成本成正比。

ZigZag Arquitectura, expands its inner void, in this case in a more private way, to search for that relational dimension that seems possible only within the block or the urban insertion and not outside the building.

But in the case of the Monte Elbruz Building there is an openness to the city even though the court has been built as a deep cut between the two blocks, creating a direct visual relationship with the street. The void inside the building is finished in itself, divided into smaller patios that serve the ground floor apartments. These do not assume a public role, but are clearly at the service of each residence. The facades of the building, in which a transparency prevails that works to drain the block, are dynamic and open, compensating for the absence of external relations.

In this sense, the three projects discussed thus far, as well as that of Frédéric Schlachet Architecte to follow, are illustrative of the present structure of the contemporary city, which is gradually losing the concept of public open spaces as primary meeting places and sites for relationships between individuals. These roles are left, according to logics which grow ever stronger, to other spaces: shopping malls and museums, for example. Consequently, the internal courtyards of city blocks no longer have the sole purpose of ensuring conditions of light and health, but have become private places with a social dimension or, where feasible, a portal between the building and the city, as in the two previous cases. These are places that are not fully public, allowing themselves to be crossed only partially by the flows of the city or, indeed, deciding what degree of violation they will allow. The concept of the collective residential building, including its social decline, is deeply influenced by this condition of contemporary life.

**Social Housing Complex, Paris_Frédéric Schlachet Architecte**

We find the same logic and the same interaction between the built space and the private courtyard in the Social Housing Complex by Frédéric Schlachet Architecte. In this project of the French architect there is also an interesting development in terms of the insertion of its architecture into the block in which it is located.

The complex creates an efficient dynamic in the volumes around the inner courtyard and faces a second open space, always contained in the block to which it belongs. The architect explains: "The buildings are directed onto two gardens: one serving the general organisation and layout of the plot as a whole, the other ensuring an opening (including the right to overlook neighbouring properties) and breaking from the pattern of gable walls that the other buildings present." It is a harder work than the previous ones: the volume is made of prefabricated concrete slabs, and even the interior features a hard, minimalist style that lends a sharp charm to the whole building.

在Monte Elbruz公寓顶楼俯瞰A单元

**第二部分：形式实验**

Siloetten住宅大厦——C. F. Møller Architects

C. F. Møller Architects的住宅大厦重新诠释了高楼的主题——奥胡斯市北部的Løgten小镇包括现有筒仓的新建筑。该项目以此方式在丹麦的很多城市中成为具有里程碑式的典型项目：废弃的工业筒仓。在这里筒仓被转变成中央核心筒，其中设置楼梯和电梯。住宅楼围绕着核心筒建造，突出了基础固体建筑的堆积性和滑动性，就像由乐高积木砖块建造成的一样。最后形成了一个动态系统，在这个系统中上空空间和洞口占据了整座高楼，使建筑明亮活泼起来，为四周的景观提供了一个开阔的视野。该设计代表了高层建筑理念的进步，将建筑转变为一座垂直的景观，而这种观念是在不断地变化发展的。筒仓的象征性作用已在C. F. Møller Architects的高楼里呈现出来了，现在它已成为一个新的城市坐标。在水平方向占主导地位的城市景观中，新的建筑采用了垂直的尺寸和密度，提供了一个在地区范围内延伸的可能性，并大大地改变了建筑结构和土地使用的比例。C. F. Møller Architects的高楼展现了这些特点，并被称为“乡村高楼”。

木料场社会住宅——O’ Donnell+Tuomey

正如C. F. Møller Architects的草案一样，由O’ Donnell+Tuomey设计的木料场社会住宅将空间、传统和革新紧密地联系了起来。实际上，就立面材料来讲，爱尔兰建筑师的项目与城市景观存在一种连续性，但是就建筑的形式策略而言，该项目在建筑层面正在尝试新的解决方案，因为它在都柏林的历史结构中通过新基地的规划设计与城市规模产生了相互作用。

这座有着47间寓所和一个街道层公共活动室的建筑，借助于盎格鲁—撒克逊等国家的传统材料——红砖作为外层，使其外观连贯一体，但它由建筑的线性序列所界定，其中设有多个带有落地窗的大型洞口。上空空间和洞口的节奏是不规律的、多变的，起到提示现代城市生活产生的干扰的作用。O’ Donnell+Tuomey很好地平衡了创新与传统，从未向地方建筑语言或过去传统的建筑语言屈服。尽管该建筑按历史结构建造，却把形式上的参考建筑转变成了现代的风格。这样一来，老式屋顶的斜坡就激发了新式建筑的更大规模的建造。

O’ Donnell+Tuomey的建筑综合体提案只有在解读城市时才能

Unlike the other works here discussed, Schlachet's project seems to be tweaking a relationship with costs, perhaps facing a more limited budget. But the choices made do not affect the aesthetic quality of the building, which, in fact, employs a language both brutal and immediate. Neither are many concessions allowed when it comes to the courtyard. There is, however, no loss of quality and clarity in the definition of the open space. Among the social housing projects presented here, Schlachet's is the one that best represents the ability of the designer to build in a simple and effective way, confirming that quality in construction is not always proportional to cost.

**Part two: Experiments in Form**

Siloetten_C. F. Møller Architects

C. F. Møller Architects' apartment building reinterprets the theme of the tower – a new building subsuming existing silos in the city of Løgten north of Aarhus. In this way the project takes on a landmark typical of many cities in Denmark: disused industrial silos. Here the silo is transformed into the central core within which are located the vertical displacements: ladders and elevators. Around the core grows the residential building, characterized by a stacking and slipping of elementary solids, as if it were built of Lego bricks. The result is a dynamic system in which the voids and openings erode the tower and lighten the building, which opens up to the surrounding landscape. This project constitutes a conceptual evolution of the high-rise, turning it into a vertical landscape, the perception of which is constantly changing. The symbolic role of the silos is now assumed by the tower of C. F. Møller Architects, which has now become the new urban landmark. In an urban landscape predominantly horizontal, the new building introduces vertical size and density, offering itself as an alternative to sprawl and widespread employment of the territory, and profoundly changing the ratio between the architecture and land use. All these features give rise to this tower by C. F. Møller Architects being defined as a "rural high-rise".

Timberyard Social Housing_O'Donnell+Tuomey

Just as in the C. F. Møller Architects' draft, the Timberyard Social Housing by O'Donnell + Tuomey creates a profound relationship between place, tradition and innovation. In fact, the Irish architects' project is in continuity with the urban landscape with respect to the materials of the facade, but it experiments with new solutions at the architectural level in terms of its formal strategy, as it plays with the urban dimension through its new site plan design within the historic fabric of Dublin.

The building, which contains 47 dwellings and a street-level community room, gives the appearance of continuity with the existing thanks to its surfaces of red brick, a material traditionally used

位于Løgten的新公寓——Siloetten住宅大厦，奥尔胡斯市北部

充分感觉到，因为建筑群的新设置与历史建筑形成了互补，重新思考建筑的逻辑性，但并不是单纯地复制建筑周围实地与空地的关系。该建筑群设定了新的边缘，形成一个三角形广场，这样就产生了一个半间隙的空间，该空间可用于观看城市和这个新的建筑群。

香草园住宅——Caramel Architekten

建筑与形式的实验使Carmel Architekten与统一的样式和类型学拉开了距离，因此他们在维也纳的香草园住宅呈现出了复杂和不规则的轮廓，以墙上的电缆线上悬挂着檐板为特色，而且挤出的体量和深深的露台能够使实体的严格性变得柔和。这些部分附加物整合在一起使整座建筑向外迸发，在不同高度上与各种户外区域形成互补。这座混合大楼的设计灵感来源于建筑师们创建高水准的个性化公寓的意愿，主要是有楼梯结构的复式单元，这些复式单元从立面上突出了其内部的功能。因此，该建筑有一个混合的策略模式："凉廊代表客厅，在不同结构中安装的较小窗户代表卧室。"建筑师解释道。因此建筑的前面是一个过程的结果，而不是平衡两个部分的混合体，而且建筑的前面为该程序提供了一种解读方式，使内部功能与其外观一样清楚明显，掩饰其立面的二维结构。外观的统一性——一种令人啼笑皆非的芥末黄，更清晰地显示了结构与悬垂的二维表层之间的连续性，几乎像外部表层一样为显示其功能而产生变化。

Marco Atzori以优异的成绩毕业于卡利亚里土木工程与建筑学院，获卡利亚里大学博士学位，著有专题论文"地面工程"。Marco Atzori曾在巴塞罗那和卡利亚里教授本科设计和研究生课程（"智能海岸"，巴塞罗那）。自2009年以来，与阿尔盖罗大学合作。他的作品刊登在《建筑学报》上。他经常就当代建筑学举办一些文化活动、展览及讲座。从2005年开始，他与卡利亚里的Michele Zara共同创办了Atzori+Zara事务所。

**Marco Atzori** graduated cum laude in civil engineering and architecture in Cagliari. Holds a Ph.D. from University of Cagliari, with a dissertation on "The Project of the Ground". Has taught undergraduate design studios and postgraduate courses in Cagliari and Barcelona ("Intelligent Coast", Barcellona). Since 2009, collaborates at University of Alghero. His writings have appeared in "Il Giornale dell'Architettura". He usually organizes cultural events, exhibitions and lectures about contemporary architecture. Since 2005, co-owner of architectural office Atzori+Zara with Michele Zara in Cagliari.

in Anglo-Saxon countries, but it is defined by linear sequences of blocks emptied by large openings that contain ceiling windows. The rhythm of voids and openings is irregular and varied, and serves notice of the contemporary intervention. O'Donnell + Tuomey effectively balance innovation and tradition, never giving in to the temptations of vernacular or nostalgia. The building, although worked into the historic fabric, transfers any formal reference to the present. In this way, the slopes of the old roofs inspire the more massive coverage of one of the new blocks.

The complexity of the proposal by O'Donnell + Tuomey can be fully perceived only in terms of its urban reading, since the new plant of the buildings patches up the historic fabric, rethinking its logic, but without naively replicating the relationship between the full and empty spaces of the surrounding blocks. The buildings set new perimeters and form a triangular square that generates a semi-interstitial space which serves the city and the new buildings that look out upon it.

**Herb Garden Housing_Caramel Architekten**

The experimentation with form distances Caramel Architekten from consolidated patterns and typologies, so that their Herb Garden Housing, in Vienna, presents a complex and irregular profile, featuring walls with detached overhanging cornices suspended by cables, and extruded volumes and deep terraces that unbalance the sternness of the solid. This composition of an annexation of parts seems to cause the building to erupt to the outside, recovering various outdoor areas at different heights. The design of this hybrid building is driven by the desire of the architects to create a high level of individualization in the apartments, mainly terrace-house-style duplex units, which underscore in their facades some of their internal functions. Thus, the building has a compositive strategy: "the loggias(eyes) indicate the living rooms; the smaller windows arranged in varying patterns mark the bedrooms," explain the architects. The front is therefore the result of a process rather than a compositional balance between the parties, and offers a reading of the program that materializes the internal function as manifest on the exterior skin, belying the facade's two-dimensionality. The processed unity of the surface, an ironic mustard yellow, shows all the more clearly the continuity between the structures and the overhanging two-dimensional surface, almost seeming, like the external skin, to change and mutate in order to express the functional program. Marco Atzori

# 米耶雷斯社会住宅

ZigZag Arquitectura

住宅单元

本案项目的住宅单元拥有城市和乡村的景观。它们具有双向定位，能带来对流通风和良好的视野。

住宅通过以卫生间为中心的方法设计，卫生间把住宅分成日间活动空间和夜间活动空间。日间活动空间正对内部广场，配有大型窗户以便利用阿斯图里亚斯宝贵的日光照射，而夜间活动空间则有更多可控开口，并配以百叶窗，面向城市。

住宅单元由总长2.60m的连续模块构成，提供一室、二室、三室和四室的户型，因此能满足户型多样化的要求。纵深1m的露台被设计成公寓在视觉上和空间上的延伸。露台可以起到外部走廊的作用，它由厚木板构成，构成建筑整体内的外围空间。

建筑内安装了可以集中供热和独立分配的地热系统，使用简单的家庭通用系统以及其他系统，可以对地热系统进行控制。建筑也可以通过安装在平屋顶上的镶板吸收太阳能，可以提供烧热水所需的部分能量。地下结构体系包括一个大型基础板及一个由混凝土墙和立柱组成的结构。地上是混合体系，以标准化管状支座和钢撑臂上的单向混凝土板为基础，这种结构能使空间快速安装使用，并且更节省成本，清晰度更高。

## Mieres Social Housing

### Capture the place, connect with the valley again

One of the most interesting characteristics that you notice once you get to the place is that despite being in the middle of an urban building mass, you can find limited views of the fields and meadows in the valley located in the higher part of the surrounding mountains. Our first aim was to recover this double quality of the place, making the project urban and rural at the same time.

It was essential to model the rigid traditional urban block of seven different heights to end up building a new volume of variable heights (three to seven stories), containing the complete residential program required. The rest of the program (storage rooms, garage and facilities) was designed in a common underground basement.

We wanted the building to match the environment, voids and cuttings which allowed the view of the mountains in the empty spaces between the buildings, fragments of the Asturian landscape in the distance, enabling the sun and the air to enter the inner space at the same time.

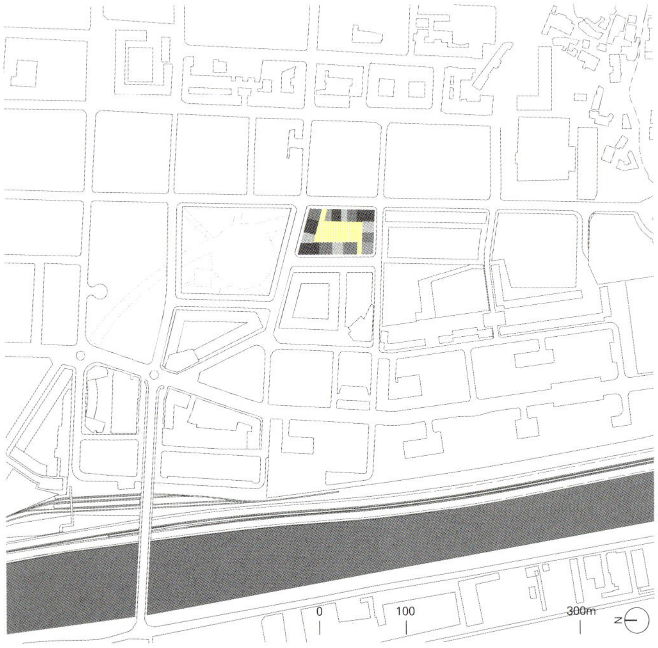

项目名称：Vivazz, Mieres Social Housing, 131 Social Dwellings
建筑师：Bernardo Angelini, David Casino
技术建筑师：Alberto López Díez
合作者：Alejandro García, Pedro Magro, André Albuquerque,
Joana Martins, Diana Jusdado, Emma Palenzuela, Manuel Sayago,
L. F.dez Bárcena
结构顾问：Jesus Hierro
设施顾问：PGI group, DYA Engineers, Adober
甲方：SEPES. Entidad Pública Empresarial del Suelo. Ministerio de Vivienda
地点：Mieres, Asturias, Spain
总建筑面积：17,840m²
竣工时间：2010
摄影师：©Roland Halbe

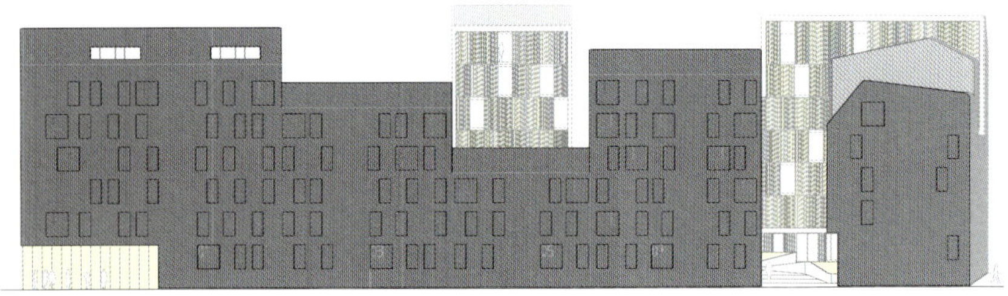

西立面　west elevation

南立面　south elevation

0　10　20m

A-A'剖面图　section A-A'

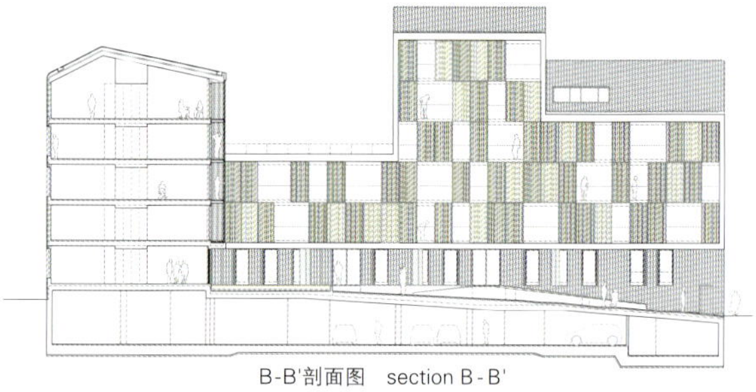

B-B'剖面图　section B-B'

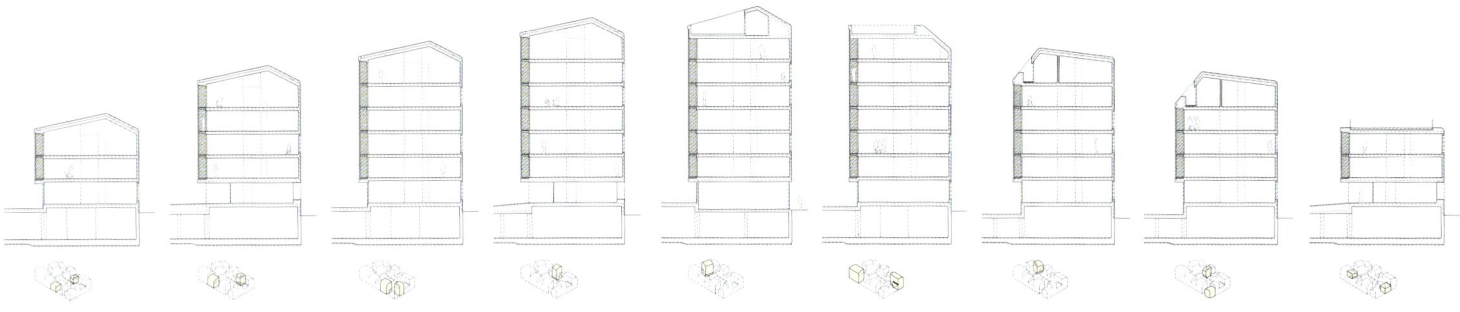

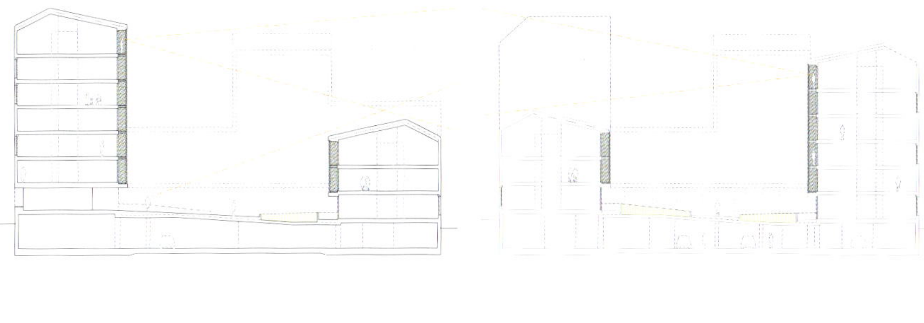

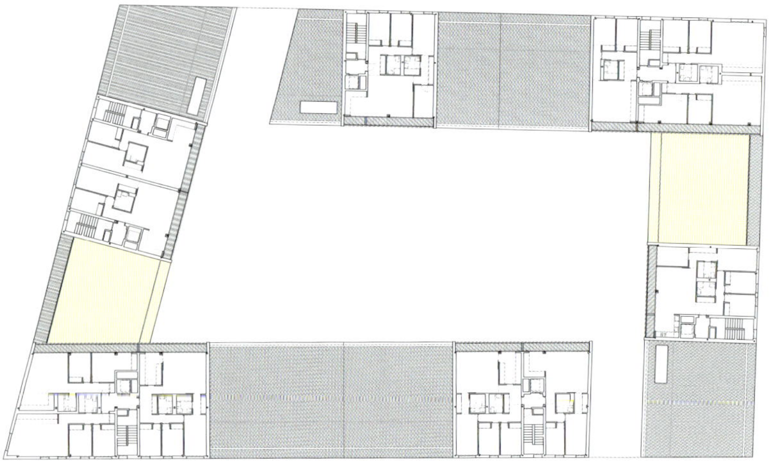

六层　sixth floor

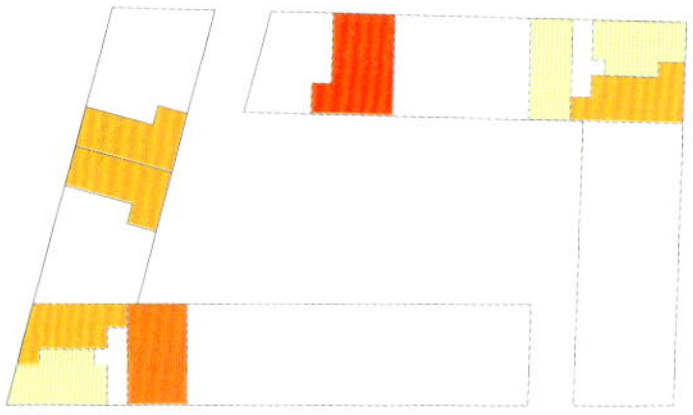

七层　level 6

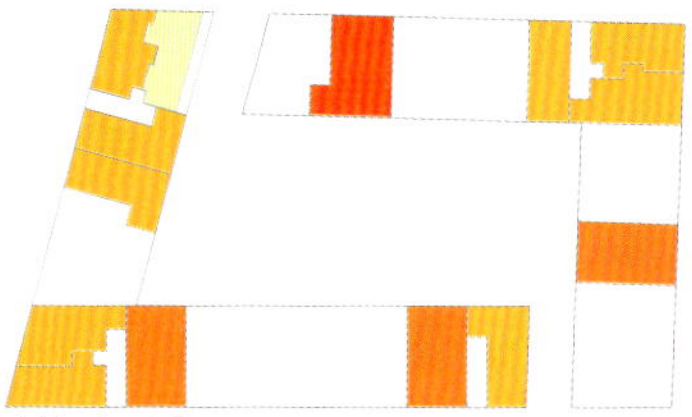

六层　level 5

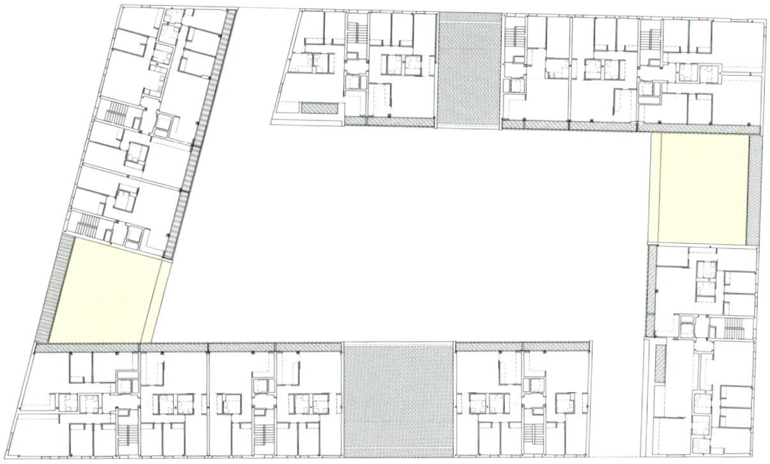

四层　fourth floor

五层　level 4

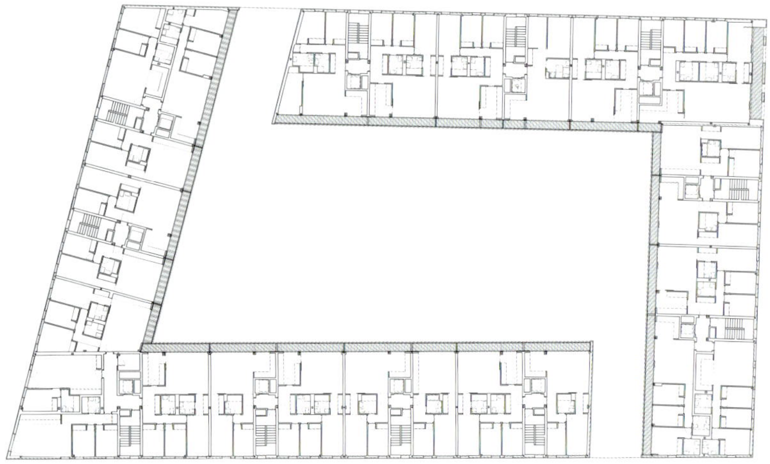

二、三层　2nd～3rd floor

四层　level 3

二、三层　level 1＋2

一层　first floor

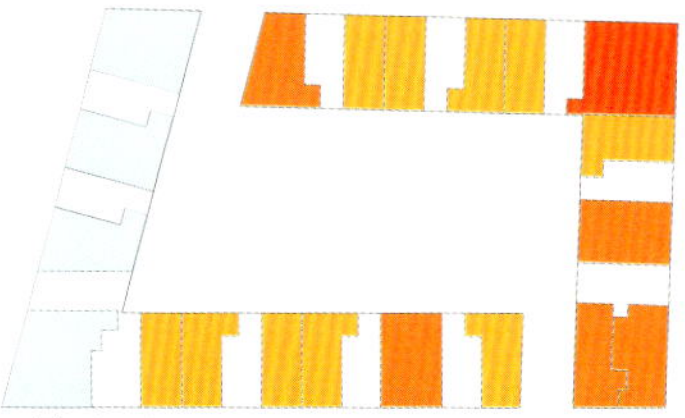

一层　level 0

户型 1
housing type 1

户型 2
housing type 2

户型 3
housing type 3

户型 4
housing type 4

商业户型
commercial

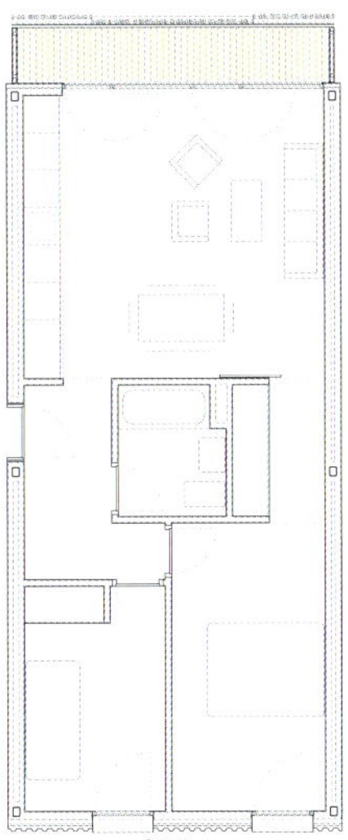
户型 1
housing type 1

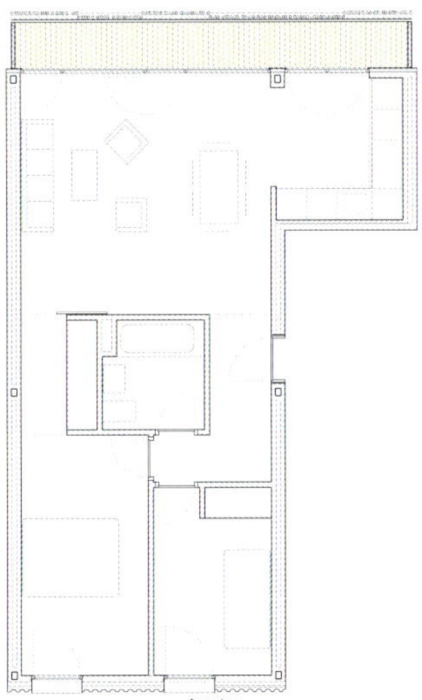
户型 2
housing type 2

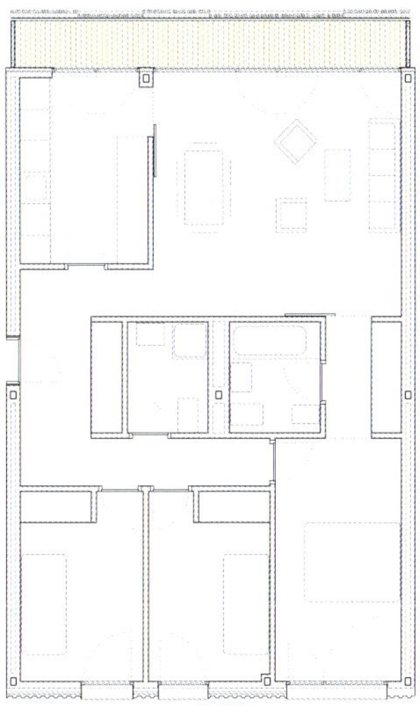
户型3
housing type 3

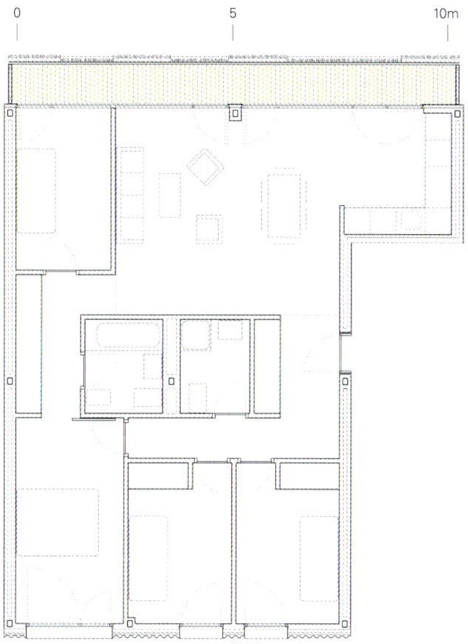

户型4
housing type 4

## An urban room, the inner plaza, a meeting place

We proposed a return to the origins of the site. An inner world that takes us back to the fields that could no longer be seen, an urban room with fragmented borders, attracting the inhabitants of Mieres through its doors, open to the city.

Opening the block was another of our priorities, and we managed to avoid turning the interior space into a forgotten back area, making it the center of the social meeting of its inhabitants as well as a place worth seeing. The day spaces of the apartments face the plaza, hence generating activity in its surrounding facades. The access was designed by breaking the block in two of its corners, creating diagonal space tension. The entrance to the apartments passes through this intermediate space, activating it and generating the necessary neighborhood relationship.

1. architectural steel facade and roof profile euromodel 44. e = 0.6mm finishing lacquer hair-supreme 35 mic. RAL color: 7022
2. horizontal substructure galvanized steel. h = 40mm
3. vertical substructure galvanized steel. h = 100mm
4. ultravent black knauf insulation glass mineral wool. e = 60mm
5. water repellent mortar filling
6. brick wall. e = 150mm
7. interior finish gypsum plaster and white smooth painting
8. wood laminate flooring
9. radiant floor system (demotic system control)
10. reinforced concrete structure slab.
11. protection plate. galvanized steel
12. window sill pre-galvanized steel plate finishing lacquer hair-supreme 35 mic.
13. pre-galvanized steel folded plate finishing lacquer hair-supreme 35 mic. RAL color: 7022
14. aluminium window bicolor (white / RAL 7022) model equi thermic by alumafel double glazing 3 + 3 / 12 / 6, with aluminium insulated blinds. RAL color: 7022
15. terrace blind steel frame formed by anular steel profiles (50 × 50 × 3) and steel platens (27 × 5) finish in anticorrosive paint and RAL color: 7022 tali wooden slats of 60 × 30mm section with exterior wood treatment
16. steel platen finished in anticorrosive paint and RAL color: 7022
17. coating aluminium panel finishing lacquer in RAL color: 7022
18. UPN-160 structural steel beam finished in anticorrosive paint and RAL color: 7022

0 0.5 1m

立面详图
elevation detail

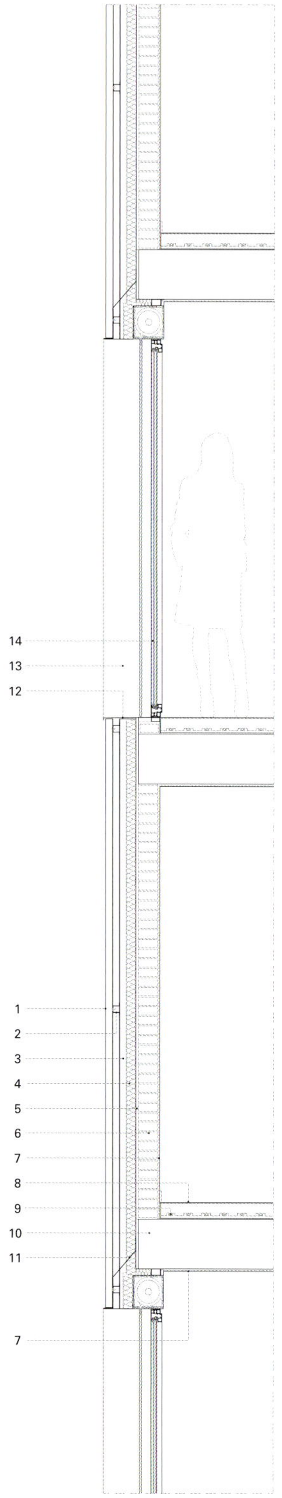


外立面详图
exterior facade detail

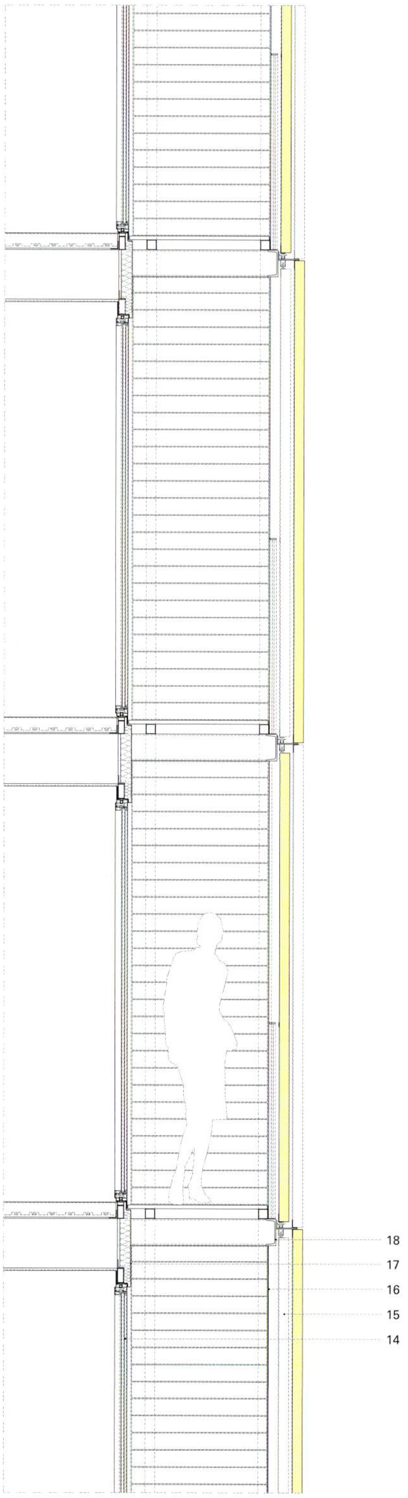


内立面详图
interior facade detail

01
03

The plaza was designed with a slight slope that absorbs the natural inclination of the site and gently links all the levels of access and main entrances. The landscaped green areas are slightly higher, containing the spaces for the exclusive use of the ground floor apartments, separating them from the public circulation areas. The space is built with concrete blocks of varying porosity, low concrete walls, which serve both as a plant container and as a bench, and native grass and groups of bamboo plants.

### Facades, dialogues with memory

The dual nature of the project results in a double material configuration. On the one hand the urban face materializes in steel and brings back the more industrial and mining image of Mieres, the former main economic engine of the city.

Once inside the building we looked for a rural reference, a return to nature and the origins of the site, the Asturian traditional porch, the use of wood, which remind us, through its vertical rhythms, of the forests of the nearby mountains.

The outer skin serves as facade and cover at the same time. It is built with panels of dark gray steel corrugated sheets, and is treated as a protective shell and as a continuous rugged casing in which the edges are rounded.

The inside of the building has a double skin, a transparent one formed by large windows that define the inner space of the apartments, and another one composed of movable wooden shutters that characterize the terraces, allowing the user to control solar radiation and have the necessary degree of privacy at all times.

### Housing units

The housing units share the urban and rural vision of the project. They have double orientation enabling cross ventilation and view.

The houses are organized by means of a central core of bathrooms that divide them into day and night spaces. The daytime areas face the inner plaza and have large windows to profit from the valuable sunlight in Asturias, while the night areas, with more controlled openings and equipped with blinds, face the city.

The housing units were generated adding up consecutive 2.60 meters modules, providing one, two, three and four bedroom units, thereby responding to the multiple types of apartments required. The terrace, one meter deep, is presented as a visual and spatial extension of the apartment. It is intended to serve as an exterior gallery made of wooden planks bordering the entire inner perimeter of the building.

A radiant floor system with centralized heat production and individual distribution has been installed, completed with a simple home demotic system which, among other features, allows control of the facility. The building also captures solar energy through the panels installed on the flat roofs to support partly the production of hot water. The structural underground system consists of a large slab of foundation and a structure of concrete walls and pillars. Above the ground, the system is mixed, based on one-way concrete slabs on standardized tubular supports and steel braces which allow faster implementation, and greater economy and clearness of the space. ZigZag Arquitectura

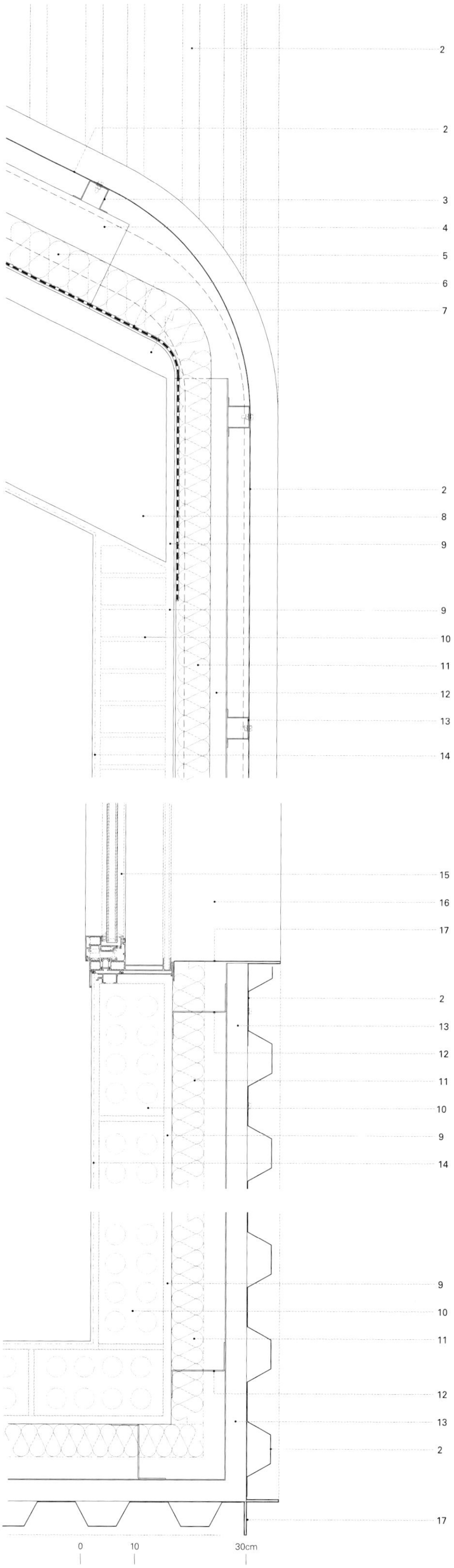

1. pre-galvanized and folded steel plate finishing lacquer hair-supreme 35 min. RAL color: 7022
2. architectural steel facade and roof profile euromodel 44. e = 0.6mm finishing lacquer hair-supreme 35 mic. RAL color: 7022
3. roof substructure galvanized steel. h = 40mm
4. roof substructure galvanized steel. h = 170mm
5. roof insulation rock wool rockplus e220. e = 80mm
6. waterproofing paint
7. concrete leveling layer
8. reinforced concrete structure slab. e = 300mm
9. water repellent mortar filling
10. brick wall. e = 150mm
11. ultravent black knauf insulation glass mineral wool. e = 60mm
12. vertical cubestructure galvanized steel. h = 100mm
13. horizontal substructure galvanized steel. h = 40mm
14. interior finish gypsum plaster and white smooth painting
15. aluminium window bicolor (white / RAL: 7022) model equis thermic by alumafel double glazing 3 + 3 / 12 / 6, with aluminium insulated blinds. RAL colour: 7022
16. window sill pre-galvanized steel plate finishing lacquer hair-supreme 35 mic. RAL color: 7022
17. pre-galvanized steel folded plate finishing lacquer hair-supreme 35 mic. RAL color: 7022

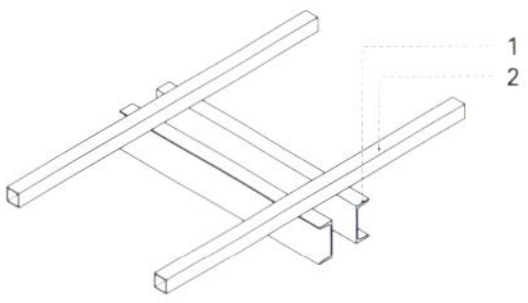

JPN–160结构钢梁
露台横向底部结构
JPN-160 structural steel beam
horizontal substructure of terrace

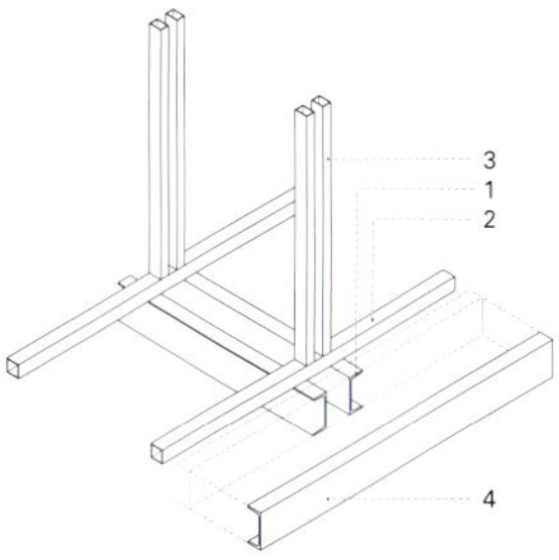

JPN–160结构钢梁
露台横向和纵向底部结构
JPN-160 structural steel beam
horizontal and vertical substructure of terrace

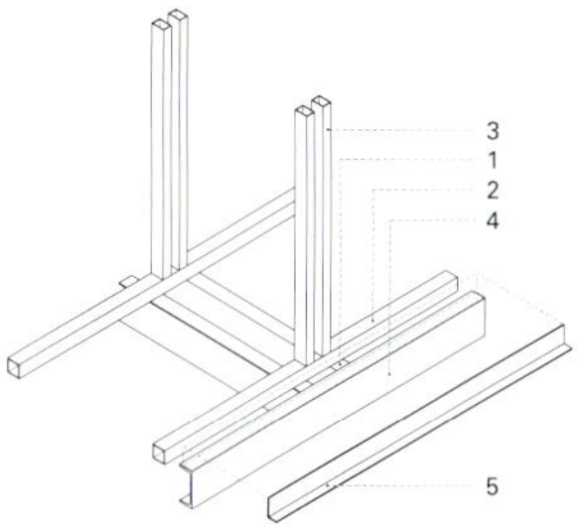

角钢型材
angular steel profile

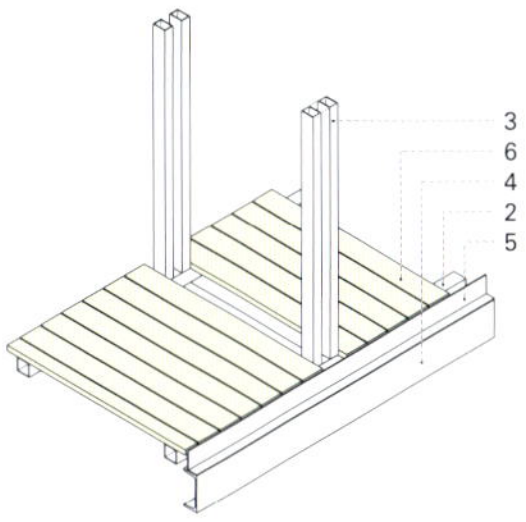

斜面厚木板
tali wooden planks

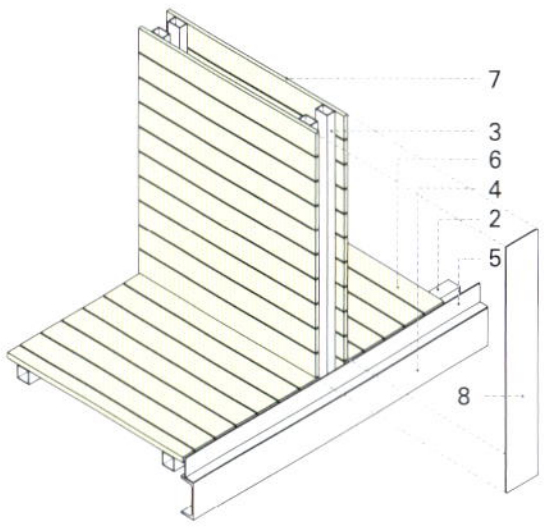

露台上的分隔板
dividing panels in terraces

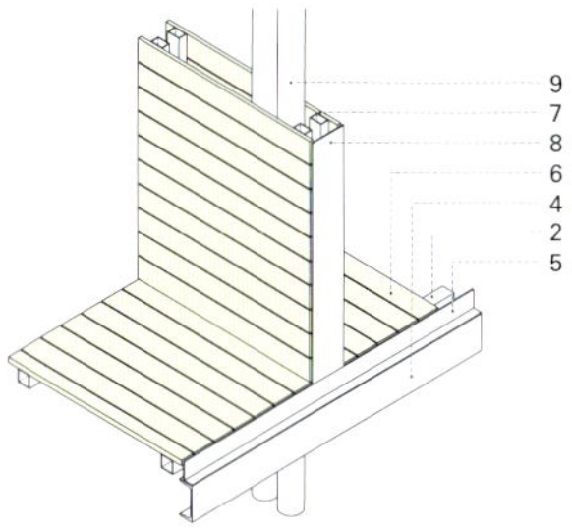

楼梯轴测详图
axonometric detail of terrace

1. UPN - 160 structural steel beam finished in anticorrosive paint and RAL colour: 7022
2. horizontal substructure of terrace steel square tube 60 × 60 × 3mm finished in anticorrosive paint and RAL colour: 7022
3. vertical substructure of terrace steel square tube 60 × 60 × 3mm finished in anticorrosive paint and RAL colour: 7022
4. IPE - 160 structural steel beam finished in anticorrosive paint and RAL color: 7022
5. angular steel porfile 85 × 45mm finished in anticorrosive paint and RAL color: 7022
6. tali wooden planks. e = 23mm with exterior wood treatment installed with hidden fasteners
7. dividing panels in terraces tali wooden planks e = 15mm with exterior wood treatment installed with hidden fasteners
8. steel platen finished in anticorrosive paint RAL color: 7022
9. drain pipe

# 社会住宅综合体，巴黎

Frédéric Schlachet Architecte

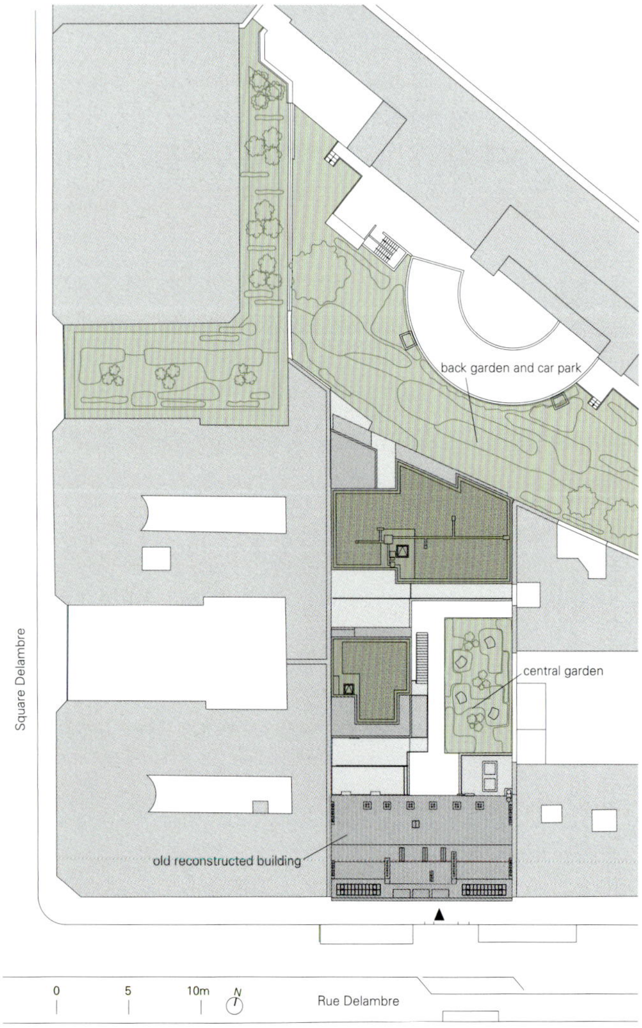


本案项目由巴黎医院组织(AP-HP)发起，其目的是优化该组织在巴黎的房地产开发，设计的项目包括商业地产和为其员工提供的住房。

在巴黎第14区的这个部分有一些废弃的地下矿场。所以，建筑的基础设在那些已经被填满的地下走廊上面25m高的立柱上。

建筑体距离马路有一段距离，处于其他建筑环绕之内的一个位置，位于一栋部分重建的原有建筑后面。建筑体巧妙地延伸出去，与附近的一座建筑共用一个地下停车场，因此不必单独设计出一条通道。它还可以俯瞰周围的建筑，这是项目体量感和空间组织的一个基本方面。

建筑体面对两个花园。一个影响了整体的布局和总体的构造，而另一个则创造了开放空间，打破了其他建筑展示的山墙模式，而且还确保了俯瞰周围建筑的视角。

一层是进行商业活动的场所和公共经营场所，构成了上面各层公寓的基础。各层的设计考虑到了项目所在地复杂的边界以及与邻近地块相关的限制条件。在这里，建筑师灵活自如地运用了体量，采用了不断变化的规模和留白空间，通过这些使各不相同的公寓具有了各自的洞口、各自的视角、各自的光线。

裸露的混凝土无论是粗糙的、石头状的还是光滑的，都无处不在。正是这种材料确保了体量的运用，而且和窗户、铁构件与光滑、抽象的百叶窗形成了对比。

在这个距离繁忙的马路和车流有一段距离的花园式环境里，去活化的混凝土无需任何防护剂。在公寓内部，地板全部采用天然材料：油地毡和瓷砖。涂料采用不含溶剂的丙烯酸树脂。

公共空间的混凝土是裸露的。由隔音材料构成的吸声天花板也是用天然纤维制成的，主要是水泥和木质纤维。

屋顶植被能够保留雨水，用来灌溉花园，同时也带来了更好的保温隔热效果。这种绿色屋顶和底层的花园相呼应，向较高层的住户展示了一种植物的连续性。

本案项目的开发特点鲜明：其一是与环境完全融合，其二是考虑到了所有与住宅项目相关的可持续发展问题。项目嵌入并通向其所在地的内部，而不是被推向所在地的边缘。这使它成为一个特例，这种特例应该成为一种规则，一种建造和发展城市可能采取的模式。在邻近的地块之间创造出新的建筑体，这在本项目中成为可能，这种设计自由度加上使用原有停车场的做法为公寓设计提供了最好的朝向，带来了最佳的视野和最多的自然光线。

## Social Housing Complex, Paris

Initiated by the Paris hospital group (AP-HP) with the aim of optimizing its property development in the city, this project was designed to include commercial properties alongside housing solutions for its staff.

There are disused quarries underground in this part of the 14th district in Paris. As a result, the foundations of the building are on 25m columns above underground galleries, which have been filled in.

The complex is set back from the road occupying an interior situation surrounded by other buildings and is situated behind an existing building which has been partially reconstructed. A clever extension means it shares an underground car park with a neighboring building thus avoiding the necessity to create a separate access. It also benefits from a right to overlook neighboring properties, which are a fundamental aspect of the volumetric feel and the spatial organization of the project.

The buildings face onto two gardens. One of them serves the general organization and layout of the plot as a whole and the other creates an opening, breaking up the pattern of gable walls that the other buildings present, and includes the right to overlook

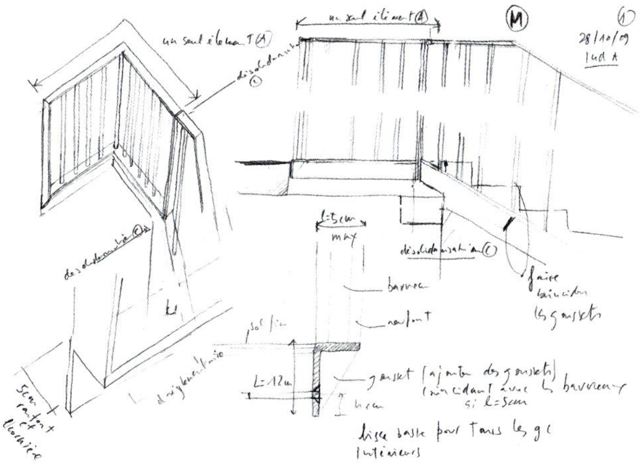

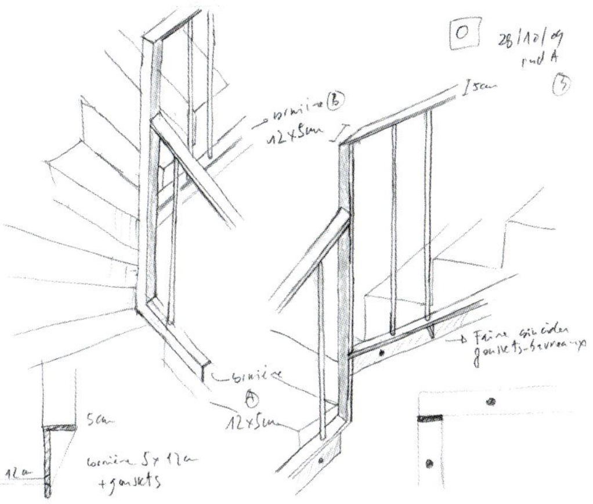

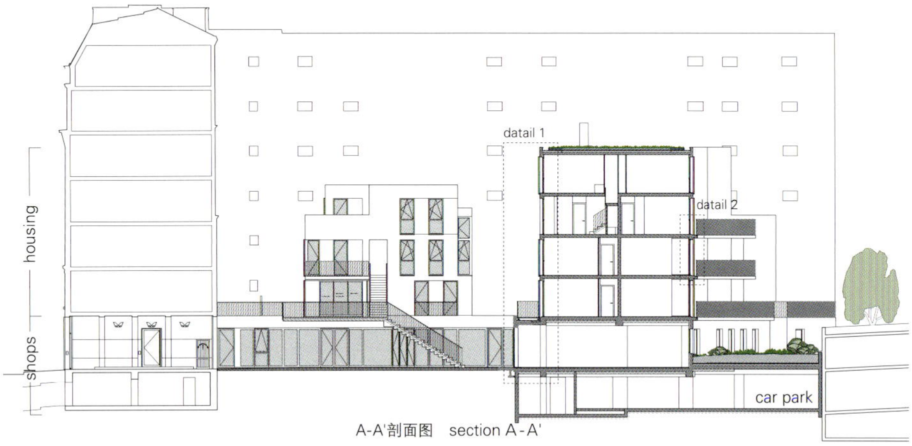

A-A'剖面图 section A-A'

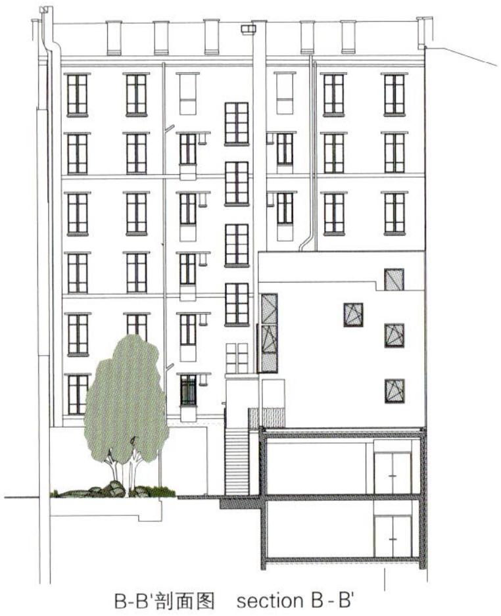
B-B'剖面图 section B-B'

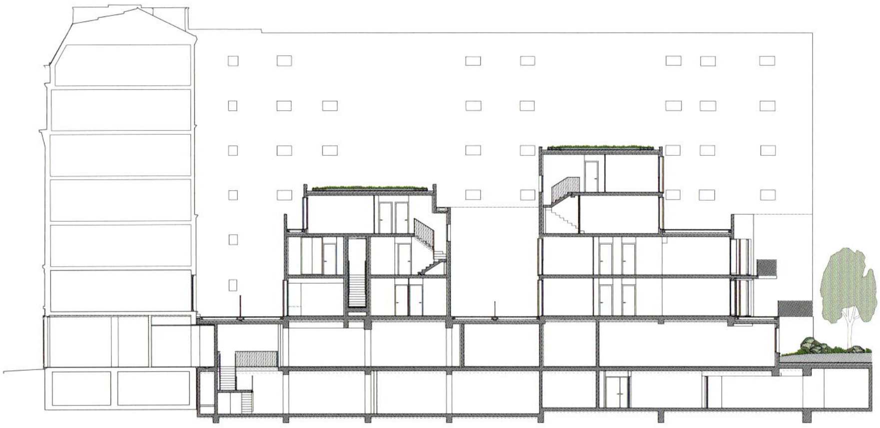
C-C'剖面图 section C-C'

北立面 north elevation

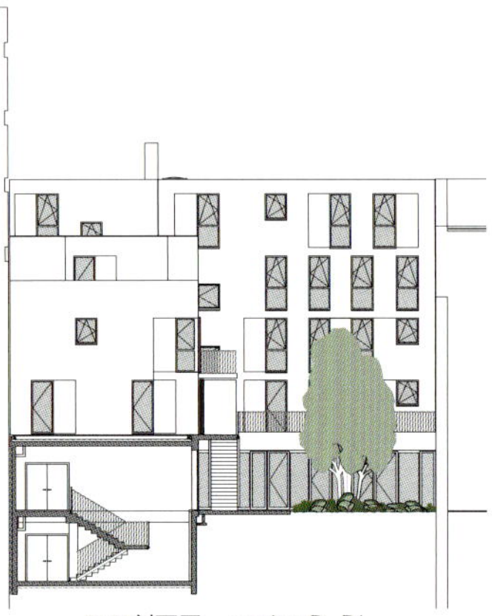
D-D'剖面图 section D-D'

neighboring properties.

The ground floor, accommodating the commercial activities and the communal premises, constitutes a base over which the apartments are laid out. Their disposition takes into account the complex boundaries of the plot and the restrictions relating to the neighboring plots. Here, the architecture plays with volumes, with changing scales and empty spaces via which the flats, each one different, find their apertures, their viewpoints, their light.

Bare concrete, whether rough, stony or smooth, is omnipresent. It is this material which allows the play of volumes and contrasts with the windows, ironwork and smooth, abstract shutters.

In this garden environment, set back from the busy roads and traffic streams, the deactivated concrete does not require any protective agents. Inside the apartments, the floor coverings are made entirely of natural products: linoleum and ceramic tiling. The paint is acrylic resin without solvents.

In the communal areas, the concrete is left bare. The soundproofing making up the acoustical ceilings is also made of natural fibers, mainly cement and wood fiber.

The vegetated roofs retain rain water, which can be used for watering the gardens and also provide better heat proofing. This type of green roof echoes the ground floor gardens and presents a vegetal continuity to the residents of the higher floors.

This development is characterized by its complete integration with its environment and by all the sustainable development considerations associated with housing projects. The buildings are implanted on and open onto the interior of the plot rather than being pushed to its edge. This makes it an exception, which should become the rule, a possible mode of constructing and intensifying cities. The freedom offered by the possibility of imagining the complex within the neighboring plots and with the existing car park has allowed the best possible orientation for the apartments providing them with maximal views and natural light.

Frédéric Schlachet Architecte

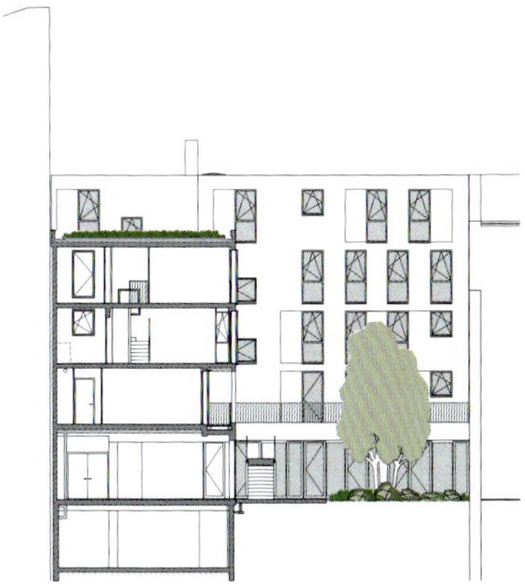

E-E'剖面图　section E - E'

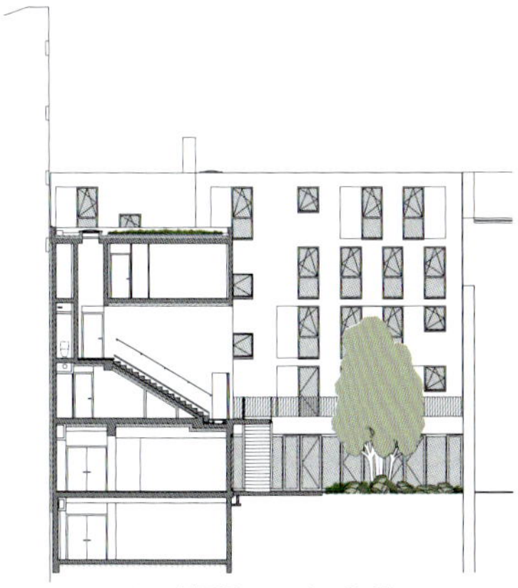

F-F'剖面图　section F - F'

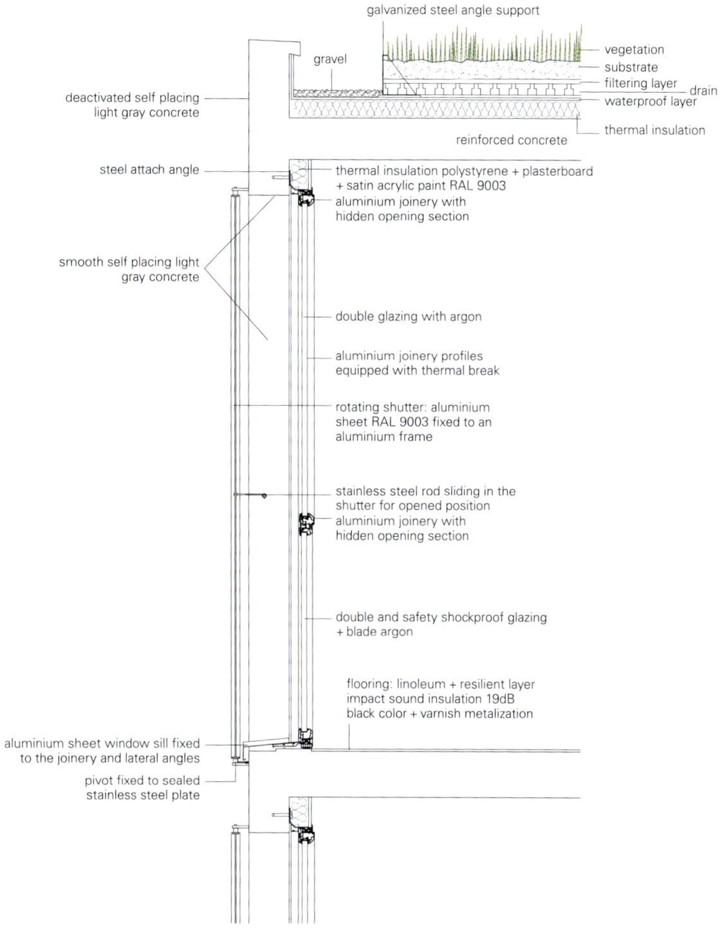


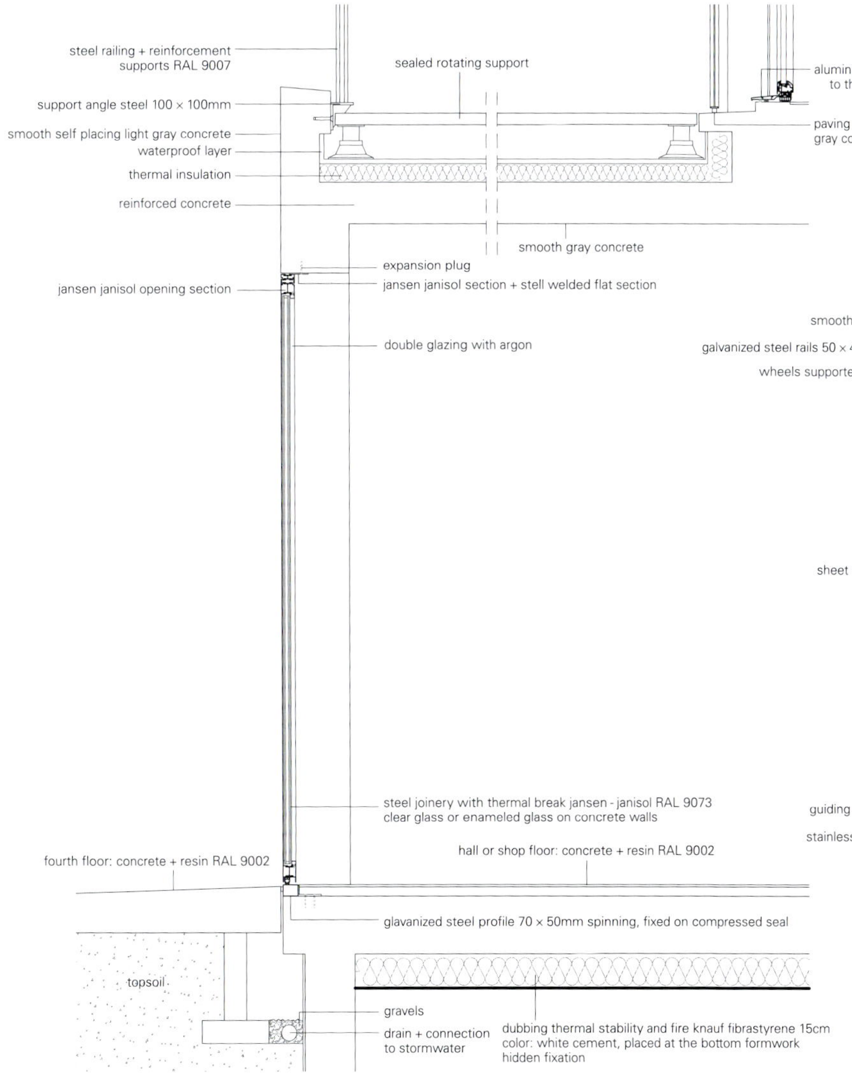


详图1　detail 1

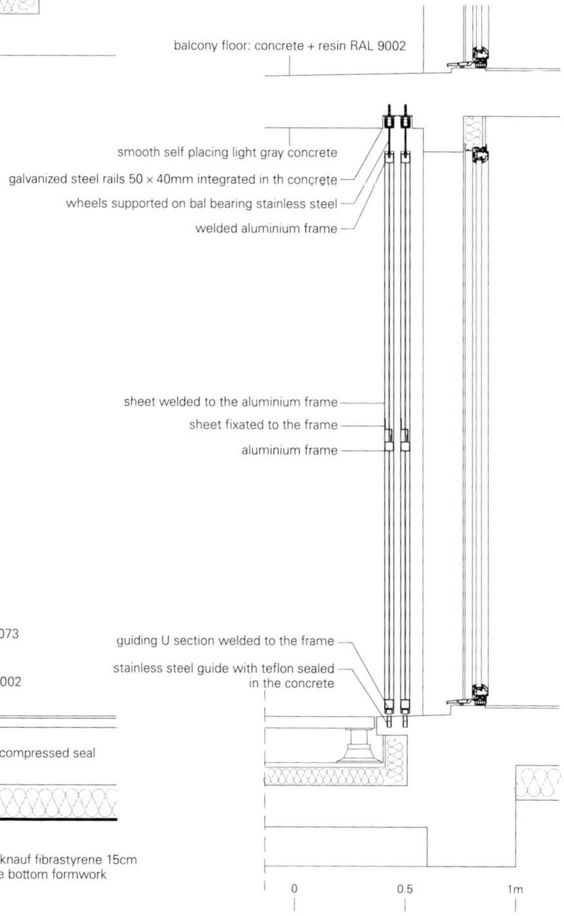


详图2　detail 2

项目名称：Social housing building and shops in Paris
建筑师：Frédéric Schlachet Architecte
甲方：SNI-SAGI, Contractor: AMT
地点：Paris, Site area: 1,650m²
造价：EUR 3,365,100
竣工时间：2010.6

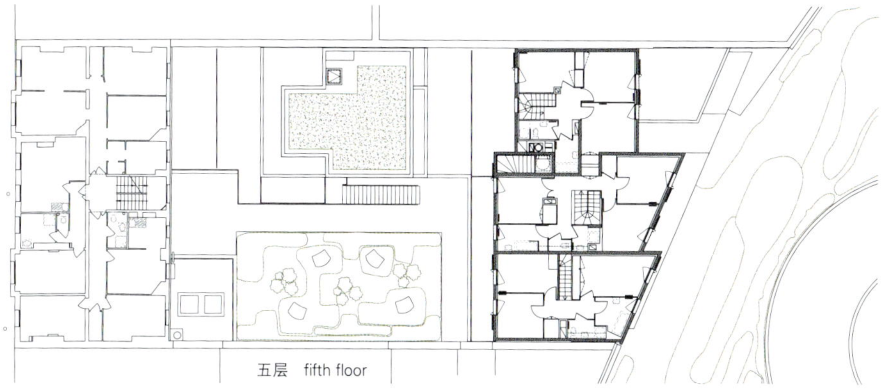

五层　fifth floor

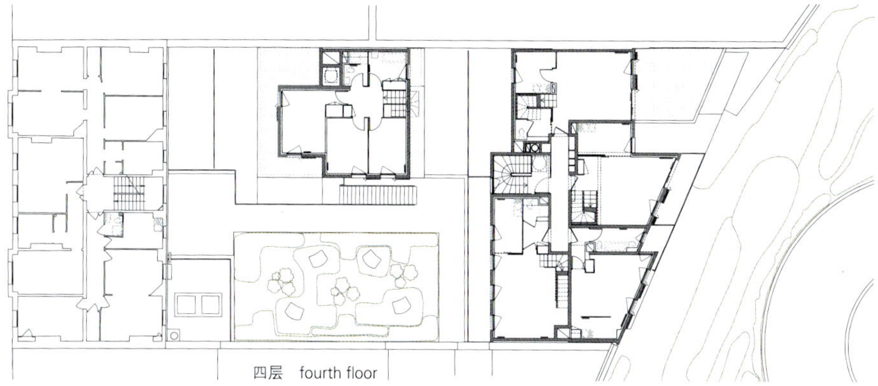

四层　fourth floor

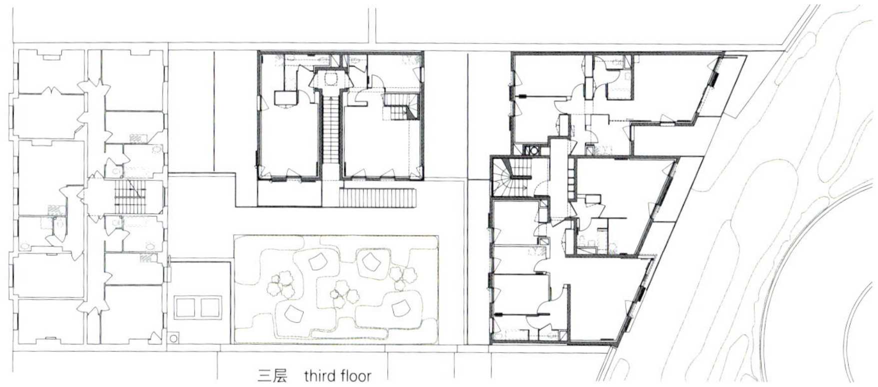

三层　third floor

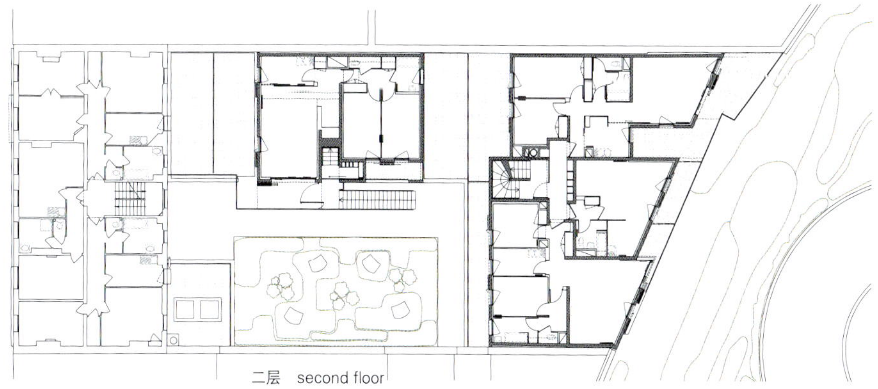

二层　second floor

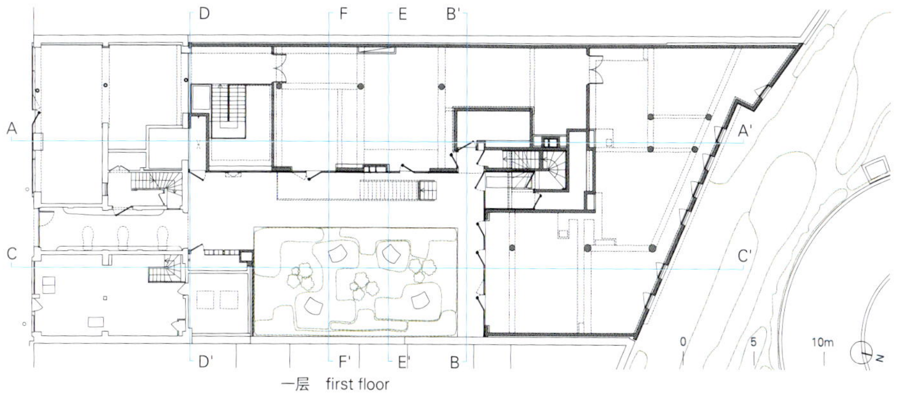


一层　first floor

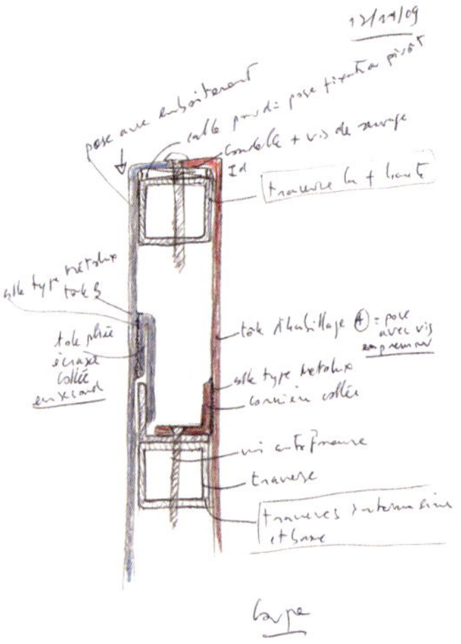


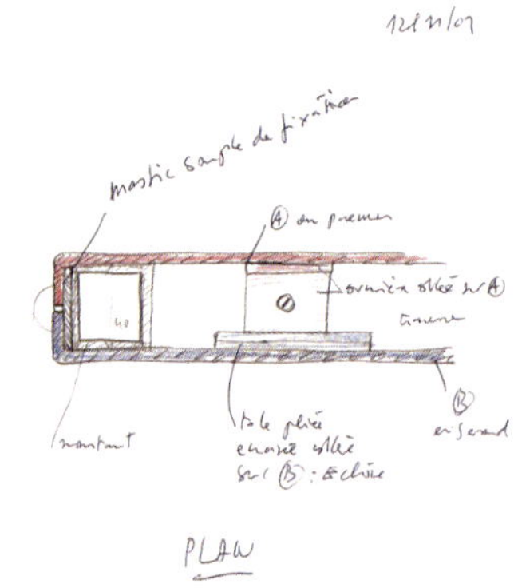


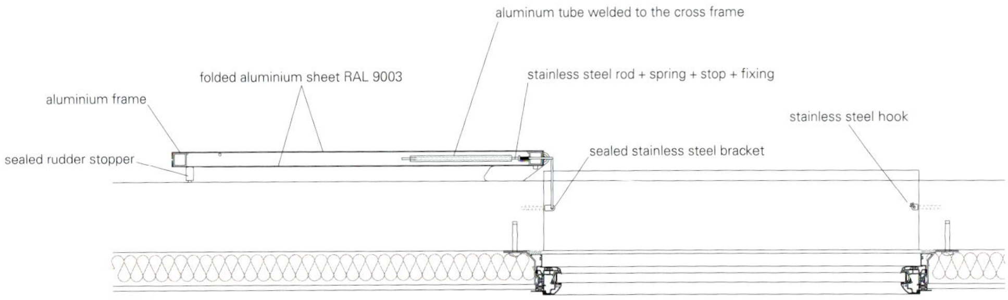


百叶窗打开时的水平剖面　open shutter horizontal section

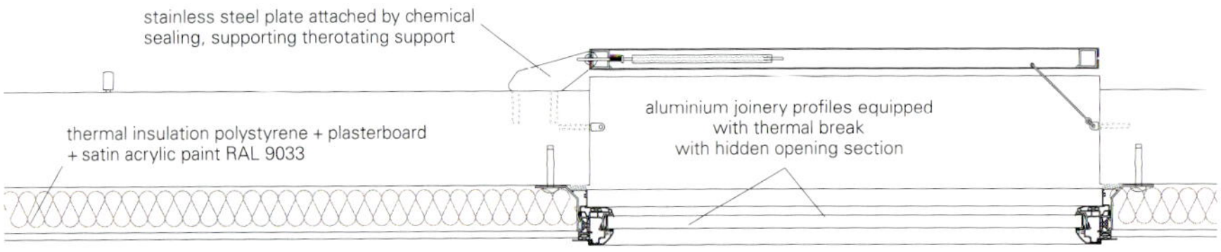


百叶窗关闭时的水平剖面　closed shutter horizontal section

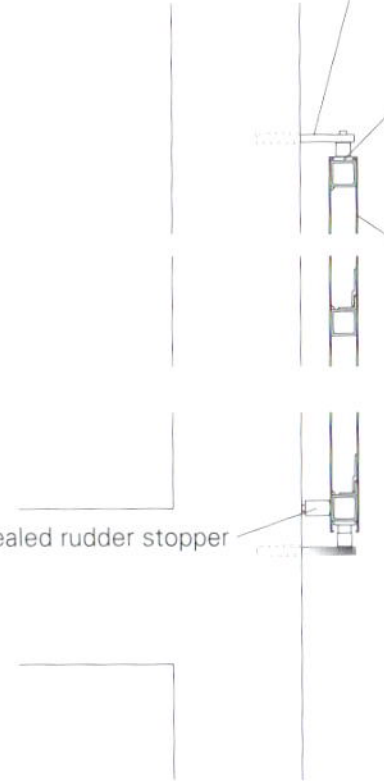


open shutter vertical section

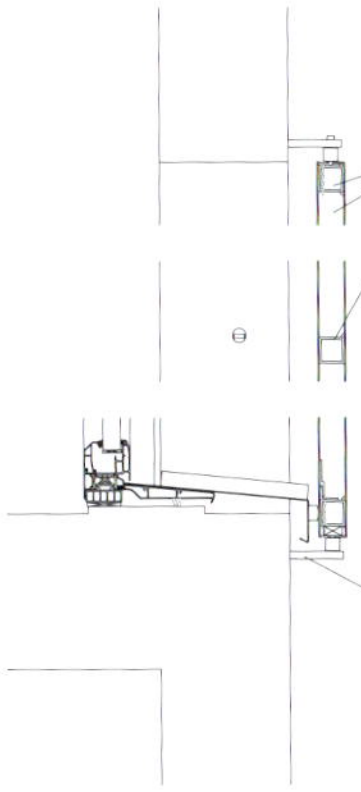


百叶窗关闭时的垂直剖面
closed shutter vertical section

# 44户社会住宅

Magén Arquitectos

本案建筑位于距萨拉戈萨50km远的陶斯特的城南边缘。

项目的出发点源于对公共空间和私人空间关系的分析。本案城市建筑按照规范而设计，将南面的原有建筑和U形布局的新建筑组织在一起。创造内部公共空间能够将新建筑与周围的原有建筑联系起来，并能创造出迷人的过渡空间。该建筑还改善了居住者的居住条件，为他们创造了可供出入、见面和聚会的室外空间。

为了处理城市建筑内部的城市化特点，建筑师采取了两种限制手段：第一是将车库设置在半地下室，其宽度与其他楼层相同，这样就不会减少公共空间；第二是将北面的一个大走廊释放出来，它能将内部空间和相邻的绿色空间联系起来。

项目还需适应其所在地的地形及规划和经济方面的限制。项目的决定性因素——材料选择上要节约以及要适应地势的坡度——通过一种抽象的语言转变成建筑的形式变化，这种转变是通过水平方向上错落有致的黑白条纹实现的。这样的构成将各个单元不同的洞口排列起来，集合在一起，在水平方向上为建筑体量带来了一定的动态效果。

至于建筑类型，则是公营企业Suelo y Vivienda de Aragón的一种公共推广形式。项目含有44户补贴住房，包括三居室或四居室的住房、车库和储藏室，建在一座U形建筑中，其中有三层住宅，车库设在半地下室。四个交通流线核心筒为每层三个住宅单元提供了出入的通道，每层都有两个朝向。

从项目设计之初到建筑内部处理，我们一直特别关注打造一个高品质的外部空间，这产生了一个令人满意的结果：在工程竣工时，中央空间与项目所在地的其他空间隔离开，融入了由街道、广场和陶斯特的公共空间以及镇议会大楼组成的网络中。

## 44 Social Housing

The building is located at the south urban edge of Tauste, which is situated 50km away from Zaragoza.

The origin of the project starts from the analysis of the connection between public and private spaces. Planned in the regulation as a city block, the organization is defined for the pre-existing building on the south and the new building, with a "U" shape plan. The definition of the interior public space is able to generate connections with the pre-existing buildings in the surrounding and to provide with attractive transition spaces. It also improves housing conditions giving them an external space of access, meeting and gathering.

项目名称：44 Social Housing in Tauste
建筑师：Jaime Magén, Fco. Javier Magén
合作者：Beatriz Olona
工程师：Rafael González
结构工程师：José Luis Sánchez
估料师：Carlos Munilla
地点：Canal de Tauste street, Tauste, Zaragoza
基地面积：1,605.72m²
建筑面积：1,167.98m²
总建筑面积：6,546.10m²
施工时间：2007~2010
造价：EUR 4,073,091
摄影师：©Jesús Granada - p.118, p.119, p.120, p.121, p.122~123, p.124[top]
©Pedro Pegenaute - p.124[bottom]

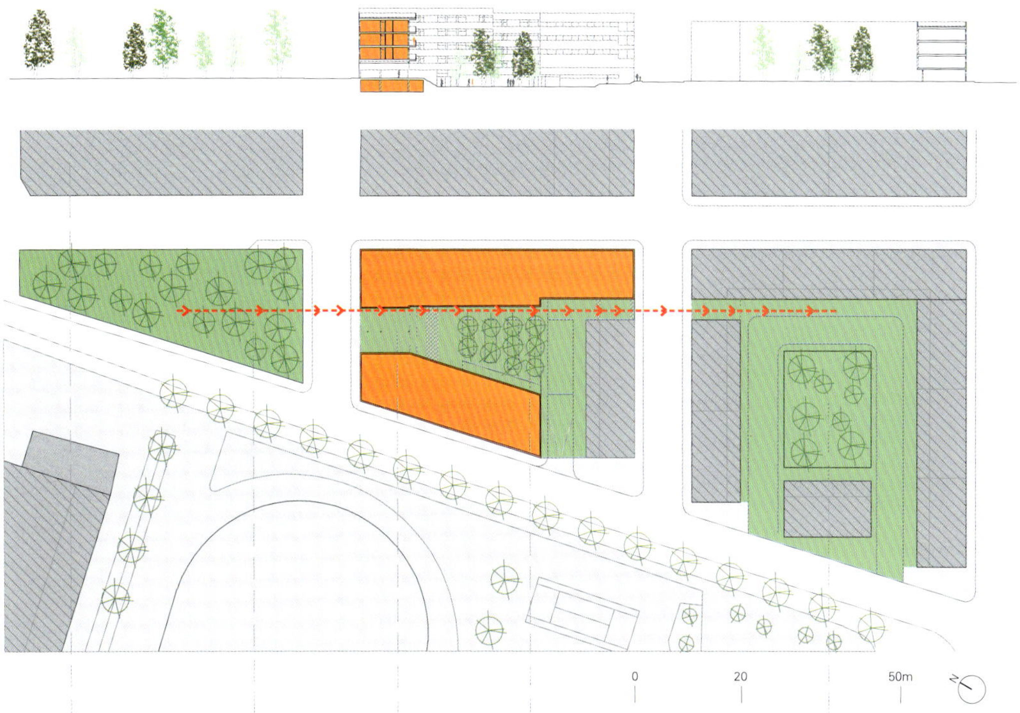

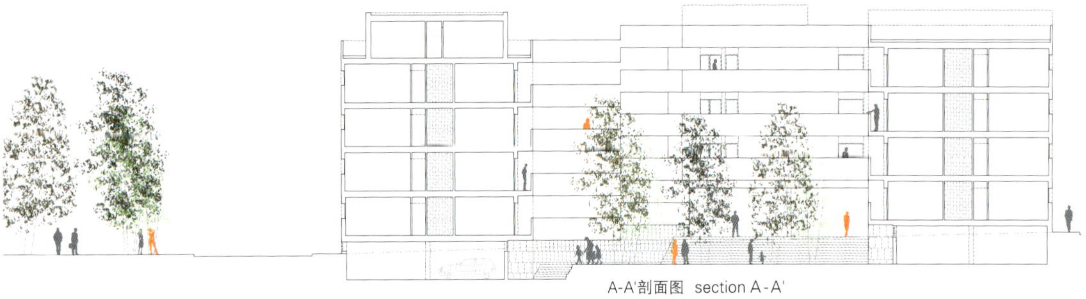

A-A'剖面图 section A-A'

B-B'剖面图 section B-B'

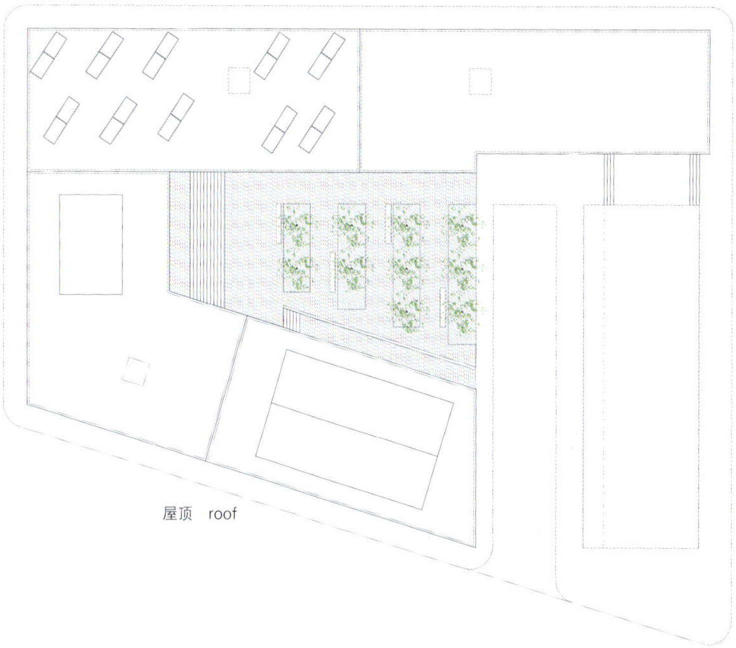

屋顶 roof

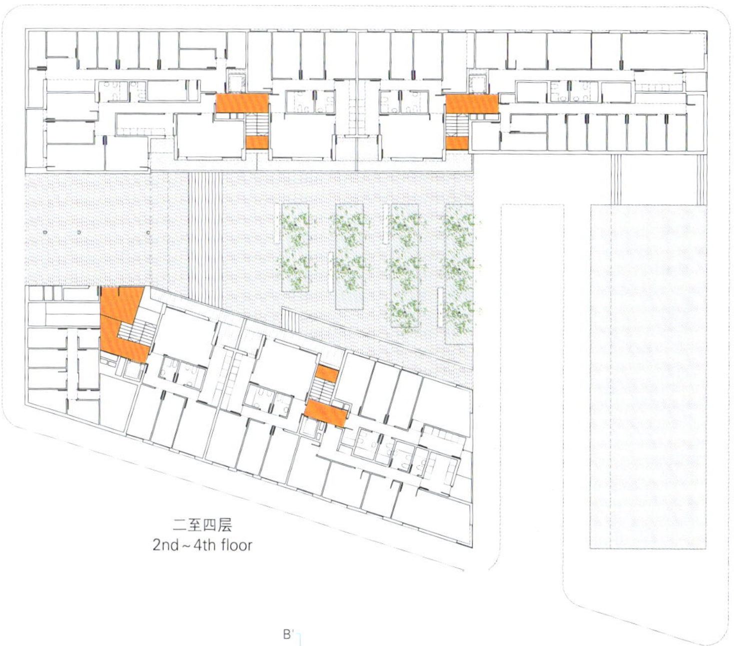

二至四层
2nd ~ 4th floor

一层 first floor

To deal with the urban character of the interior of the city block two restrictions are proposed: the first one is to fit the garage in the semibasement floor in the same width as the other floors in order not to reduce the dimensions of the public space; the second one is to free a big porch at the north side which relates the interior space with the adjacent green space.

The project also had to respond to the topography of the site and programmatic and economic limitations. The determining factors of the project – the economy in the material election and the adaptation to the slope – are converted into the formal arguments of the building by means of an abstract language based on horizontal leveled black and white strips. This composition, which arranges and brings together the different openings of the units, gives a certain horizontal dynamism to the volume of the block.

Regarding typology, this is a public promotion from the public enterprise Suelo y Vivienda de Aragón, with a program of 44 subsidized housing of 3 or 4 bedrooms, garages and storages rooms, in a block with a "U" shape plan, with three stories for the housing units and the garage located in the semibasement. The four circulation core give access to three housing units with double orientation in each floor.

The special attention given from the origin to the treatment of the interior of the block as a good quality exterior space had a happy administrative end: at the completion of the works, the central space was segregated from the rest of the site and added to the net of streets, squares and public spaces of Tauste, property of the town council. Magén Arquitectos

1. concrete pipe cap
2. concrete cleaning base. t = 10cm
3. concrete wall. t = 30cm
4. waterproofing membrane
5. drain pipe
6. extruded polystyrene joint 12 × 4cm
7. quartz polished concrete slab. t = 15cm
8. gravel layer
9. pre-stressed self-supporting hemacol slab 21 + 5
10. cement mortar levelling
11. floating platform type pergo finished with oak wood
12. suspended ceiling of plasterboard
13. suspended ceiling of heraklith wood wool panels
14. suspended ceiling of fire moistureshield plasterboard
15. formations of cellular concrete slopes
16. insulation: double extruded polystyrene panel. t = 80mm
17. geotextile sheet
18. gravel layer w min = 60mm
19. insulation of extruded polystyrene panel. t = 30mm
20. cement mortar levelling
21. pavement of concrete slabs 40 × 40cm
22. concrete slab
23. anti-skip stoneware tiles 40 × 40cm
24. preformed brick pieces. t = 10cm
25. plaster mortar single layer white scraping finished
26. collapsible steel sheet
27. Insulation of extruded polystyrene panel. t = 50mm
28. aluminium carpentry
20. dm skirting board colour white 70.10
30. collapsible aluminium sheet
31. simple partition hollow brick
32. stoneware tiling 30 × 20cm colour white
33. stoneware tile 40 × 40cm
34. white limestone cladding. t = 3cm
35. coping of galvanized steel sheet
36. concrete tile. t = 4cm
37. mortar fill

拐角住宅 corner housing
双向住宅 double-oriented housing

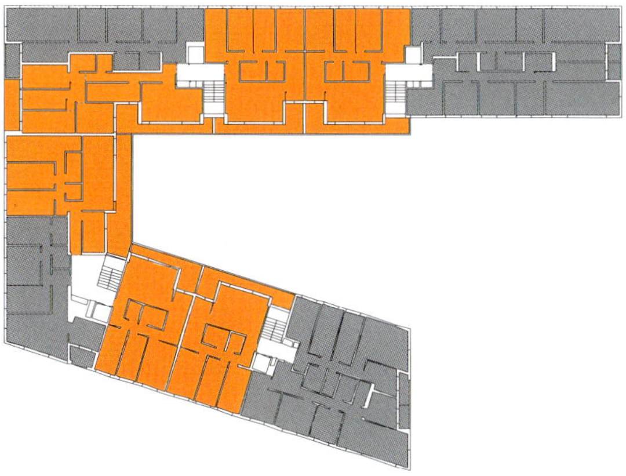

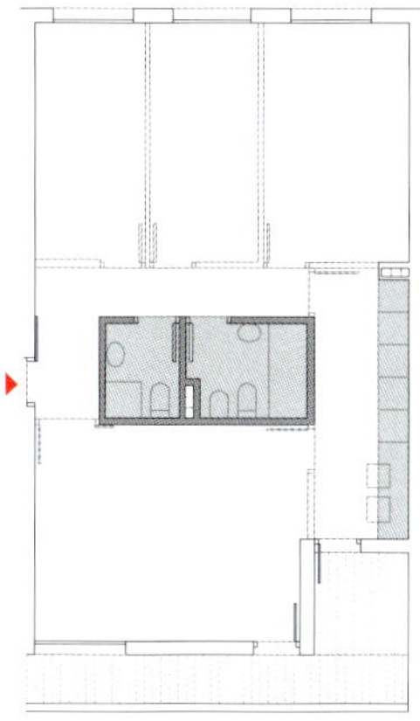

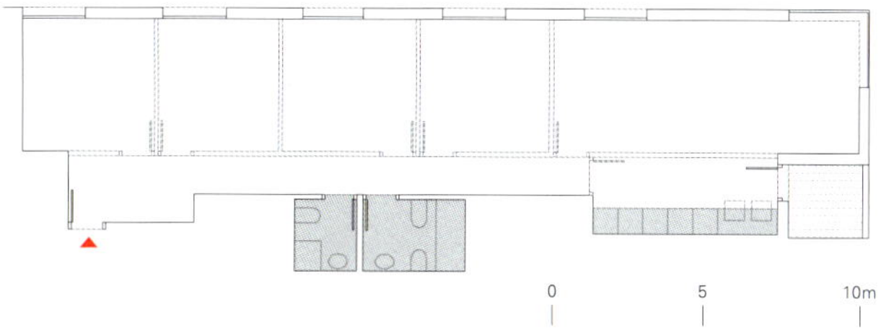

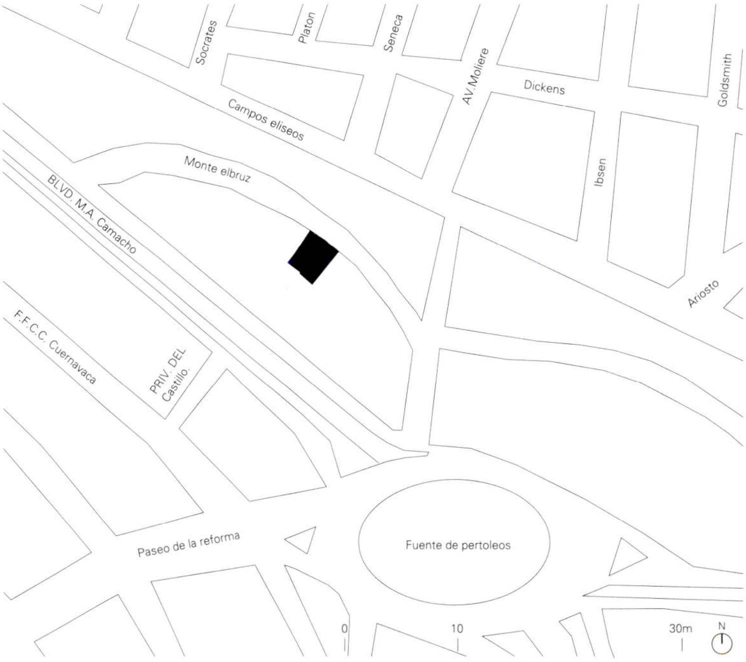


# Monte Elbruz公寓

Garduño Arquitectos

建造城中之城

项目所在地位于一个很难再开发房地产的区域，这里已被归为高密度区域，已经得到充分开发，毗邻的建筑从14到30层不等。后来建筑规范有所改变，进行项目设计时，规范要求建筑最多只能有六层，外加一个屋顶花园。

考虑到建筑密度的要求——在1300m²的基地上建造24个164m²的公寓单元，立面长30m——设计一个直线形的立面将会使60%的公寓拥有内部视野。考虑到这些变量，我们选择设计两座相对的建筑，楼间距为600cm，这样就创造出一个中央庭院作为绿化区，也能产生与项目规格成比例的立面。项目规划成功地为66%的单元提供了绿化区域。一层八个单元有私人花园，顶层八个则有屋顶花园。

每个公寓单元为164m²，包括两个卧室、三个半卫生间、工作室、客餐厅、厨房和设备间，所有这些都围绕在6m高的大厅或起居室周围。另外，所有公寓都是上下两层。项目规划还有一个公用区，包括健身室和游泳池，都在地下一层。

前面立面因为有两道狭长的、沙石色的、垂直的混凝土梁而与众不同，这两道梁通过住宅所在的体量分开。这些梁使用混凝土和碎大理石浇筑而成，然后用锤子修饰，创造出一棵树的花纹，树干始于第一道梁，树枝的痕迹扩展到第二道梁。这两道梁好像是飘浮在立面上似的，围绕建筑四周延伸，最后又回到建筑立面之前。

设计提案的主要目的之一是创造狭长的体量感，避免噪音，以一种积极的方式将项目与周围环境融合起来，并使项目与环境对比的比例恰当。

内部立面被设计成城中之城，力求通过外露的交通流线和包含玻璃框架与倾斜通风口的铝板来实现居住者的移动和互动。为了有人路过时保护每个单元的隐私，我们抬高了一层公寓的地平面，高于外面的道路90cm，而且加了一个70cm的内部矮墙，它能防止走在路上的人直接看到住宅内部。这是通过直通每个公寓的独立楼梯来实现的。

实现比例合适的内部视野后，我们的下一个任务就是用金属板覆盖相邻的立面。

雨伞的故事

“有一天，我和两三个朋友在飞行场地发现了这把破旧的雨伞。风不是很大，所以我们根本无法让伞飞起来。我们百无聊赖，所以决定试试，看能不能使这把伞飞起来。我们四处搜寻，找到了一块硬纸板做方向舵，然后在伞把上捆上一大块沥青做压舱物。我们用它试飞了几次，飞得还不错。基本上只是在升降机里拿着它走到悬崖边上，然后轻轻推一下。它能向上飞，但最后还是掉回旋转器里，不过有几次稍好一些，能飞30~40秒。”

——Aline Shkurovich

在建筑后面，我们在电梯间设计了一个纵向开口，用天空的形象覆盖起来，正是一群孩子发明了一把飞翔的雨伞的故事中那个天空的形象。

## Monte Elbruz Building

Building city within a city

The lot was located in a difficult area, originally classified as high-density and was already fully developed, adjoining buildings 14 to 30 stories high. The regulations were later changed and, when the project was being designed, the limitation considered a maximum of six levels plus a roof garden.

Given the density requirements – 24 units, 164 square meters, a 1,300 square meters surface and 30m long front – designing a linear facade would lead to 60 percent of the apartments having an interior view. Taking these variables into account, we opted for designing two side blocks within a 600cm distance of the borders, thus generating a central courtyard to play the role of the green area, but that would also generate facades proportioned in terms of the project's scale. This program succeeded in providing 66 percent of the units with a green area. The first eight with a private garden and the last eight with a roof garden.

The individual program consists of 164 square meters, with two

bedrooms, 3 and a half bathrooms, studio, dining-living room, kitchen, and service quarters, all surrounding an interior, six meter-high hall/family room. In addition, all apartments are two-leveled. The program is completed by a common area, with gym and swimming pool at the basement 1 level.

The front facade boasts two slim vertical concrete sand-colored beams separated by the volume containing the home. These pieces were cast with a double concrete face and marble grain. They were later finished with a hammer, bringing out a tree that begins with the trunk at the level of the first beam and continues the trace in the second one. These beams, that seem to float from the facade, become a parapet present throughout the building's perimeter and return to the front under the form of a slab.

One of the main goals of this proposal is to generate the illusion of slimmer volumes, avoid noise and incorporate and proportion the project with its environment in a positive fashion.

The inner facade was designed as a city within a city, striving to achieve the residents' movement and interaction by means of exposed circulations and aluminum panels containing glass frames and tilting vents. In order to achieve privacy when walking past each unit, we elevated the ground levels of each apartment by 90cm with regard to the hallways and added a 70cm interior parapet, which prevents any direct visual contact from the access hallway. This was achieved by creating independent stairs that reach each apartment.

Once having achieved proportioned inner views, our next task consisted in covering the adjoining facades with metal sheets.

**Umbrella story**

*"A couple of buddies and I found this old broken umbrella at the flying site one day. The wind was not blowing very hard so we couldn't fly. We were bored stiff, so we decided to see if we could make it fly. We scrounged around and found a piece of cardboard for a rudder, then taped a chunk of asphalt on the handle for ballast. We actually got some pretty good flights out of it. Basically just held it out over the edge of the cliff in the lift and gave it a light push. It would climb and finally get blown back into the rotor, but a couple of the better flights were around 30-40 seconds."*

-Aline Shkurovich

At the back of the building we designed a vertical opening in the elevator cube, covering it with an image of the sky that is part of a story in which a group of children invent a flying umbrella.

Garduño Arquitectos

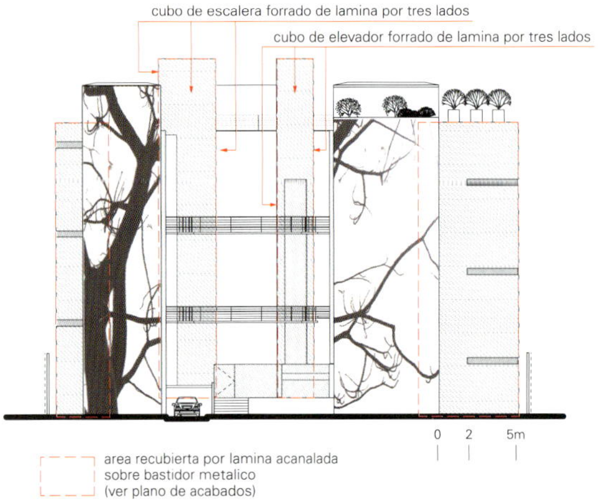


东北立面 north-east elevation

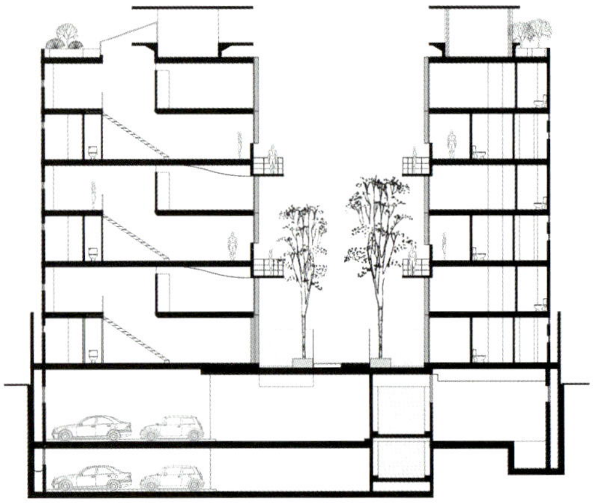

A-A'剖面图 section A-A'

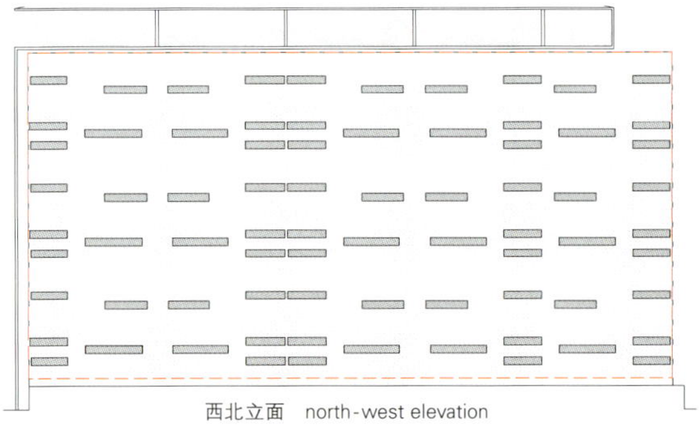

西北立面 north-west elevation

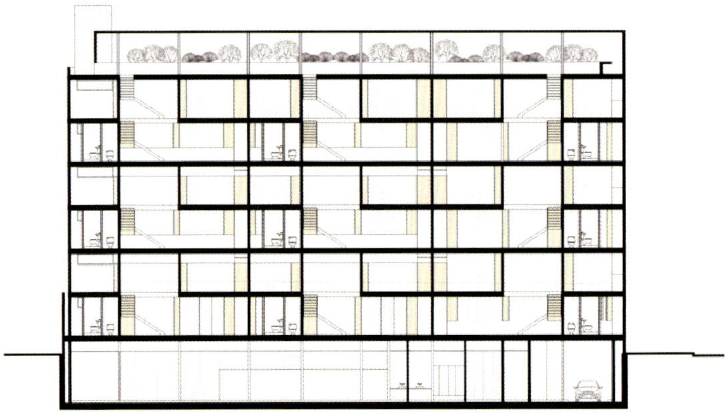

B-B'剖面图 section B-B'

项目名称：Monte Elbruz Building
建筑师：Juan Garduño, Ernesto Flores, Ricardo Guzmán, Daniel Banda
结构工程师：Diseño y Supervision S.C
电力工程师：Arturo Guerra
甲方：Bicyq Construcciones
地点：Polanco, Mexico City, Mexico
用途：multi-family house
基地面积：1,300m²
建筑面积：6,300m²
总建筑面积：800m²
建筑覆盖率：70%
总容积率：70%
建筑规模：two stories above ground, seven stories below ground
外部装饰：glass, aluminum, concrete
设计时间：2007
竣工时间：2008
摄影师：courtesy of the architect (©Sófocles Hernández)
Courtesy of the architect (©Paul Czitrom)

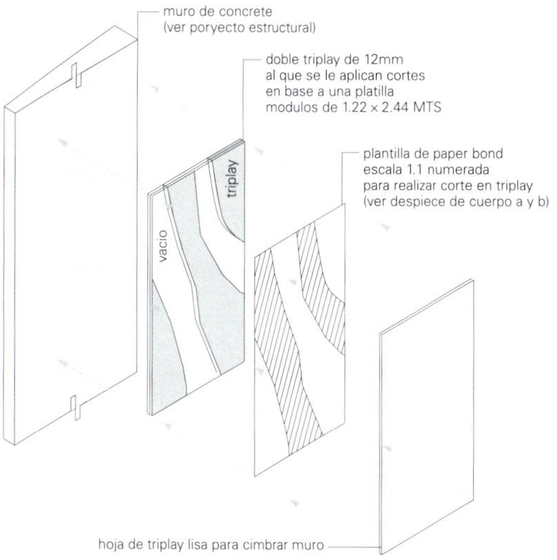


建造过程
construction process

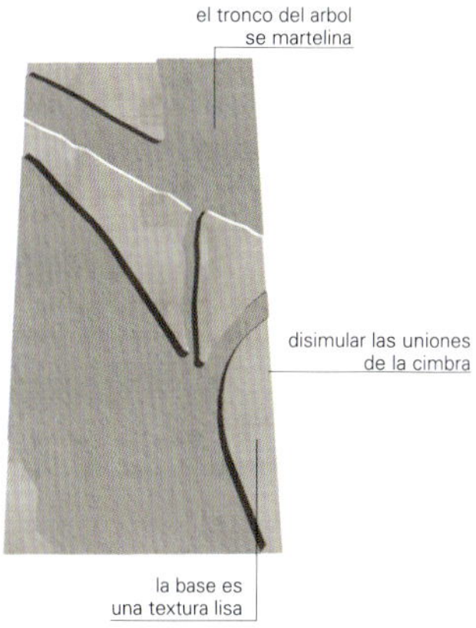


竣工模块样板
completed module sample

竣工模块 completed module

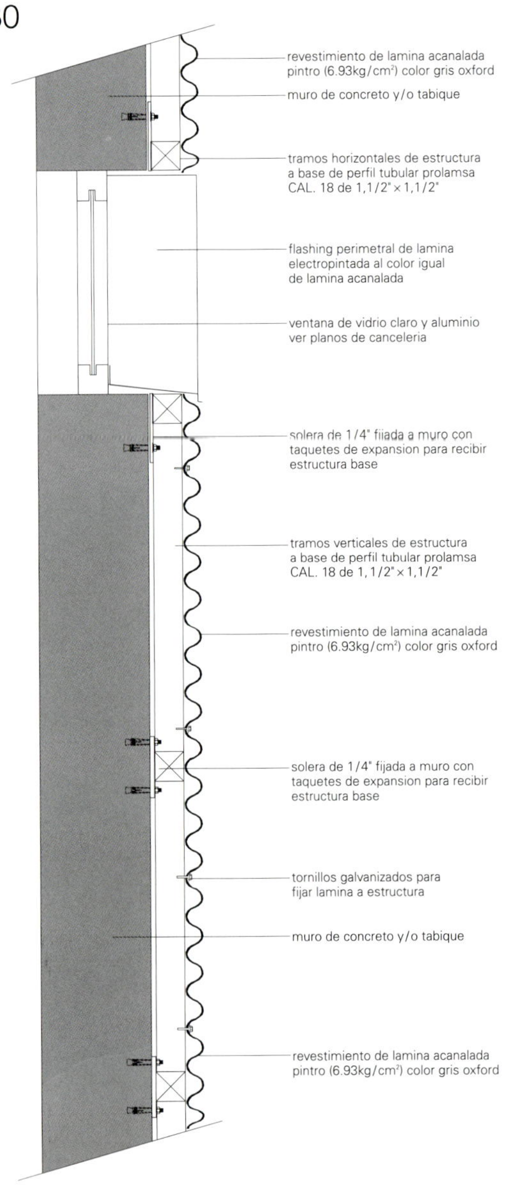

四层　fourth floor

三层　third floor

二层　second floor

1 入口
2 低层——A单元
3 低层——B单元
4 高层——A单元
5 高层——B单元
6 花园

1. entry
2. lower floor-unit A
3. lower floor-unit B
4. upper floor-unit A
5. upper floor-unit B
6. garden

一层　first floor

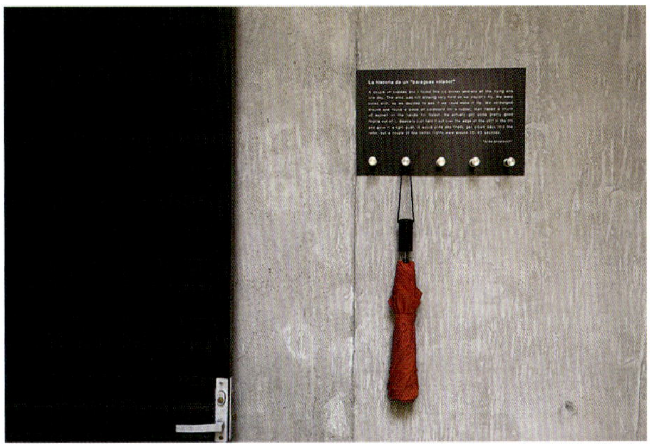

# 木料场社会住宅

O'Donnell + Tuomey

本案开发项目包括一个47套住房的新住宅计划和在都柏林这个具有历史意义的自由区的街道层建造一个社区设施。项目计划设计规模较大、特点独特的城区，将新建筑与自由区的历史特点联系起来。

都柏林市议会急于在此区创造一个样板式的社会住宅开发项目。项目概要由都柏林市议会起草，同时与当地原有居民和新居民进行了磋商。

该项目依Coombe Bypass道路工程而建。一块腹地被开放出来，城市设计的要求是要将其修建成一个新的临街面，以修复道路工程作业带来的创伤。随着原有社会住宅群的拆除和为了重新安置居民所需的新社会住宅需求的产生，一个由零售公寓构成的新城市走廊沿着科克街被开发出来。

项目包括沿着新科克街走廊设计的中等规模的六层建筑和项目基地后面较小的原有建筑。新建筑是砖石结构，配有硬木窗户、带遮阳篷的露台和屋顶花园。墙上的窗户彼此错开，与复杂的内部居住条件相呼应，同时也进一步强调了砖表面的连续性。墙体内设有凹进的嵌入式门廊和露台以及伸出的凸窗，这种设计为建筑边缘增添了深度和复杂度，同时也为住宅的私人世界和周围住宅提供了一个界面。建筑沿着弯曲的街道设置，前面种植的景观植物和阶梯为一层临街的单元带来一定的私密性。

项目的主要社交/娱乐空间是一个三角形的庭院，该庭院因为邻近公寓的住户可以看到而成为一个安全的空间。底层的靠窗座位、上层的凹进阳台及凸出的冬季花园使这个空间更具活力。

项目开发了两条新的人行路，一条穿过主庭院，一条穿过建筑东端的人造洞室，这个洞室深入到周围的城市结构中，重新构建了被Coombe Bypass道路工程破坏的城市结构。

砖和木材是项目中使用的主要材料，这与该地区原有的住宅群和工业建筑相呼应，同时也与此处曾作为木料场使用的事实相呼应。该建筑是表面覆盖砖块的现浇混凝土结构，这是一种典型的空腔结构。混凝土结构使立面上的洞口可以互相错开，同时通过垂直建造内部隔墙使公寓设计具有更高的灵活度。窗户向内凹进一整块砖的距离，让洞口看起来更深。所有洞口都有坚固的悬挂式砖砌过梁和砖砌窗台，该设计创造了一种从砖立面中穿透的感觉。庭院也铺上了砖块，与立面相称。Iroko木窗和木遮阳篷未进行处理。

## Timberyard Social Housing

The development consists of a new housing scheme comprising of 47 dwellings and a street level community facility in the historic Liberties area of Dublin. The development proposed to provide scale, identity and a piece of living city, which connects new development in the area to the historic character of the Liberties.

Dublin City Council were anxious to produce an exemplar residential social housing development in the area. The brief was set out by Dublin City Council and in consultation with the needs of the local and new residents.

The project was generated by the construction of the Coombe Bypass. A back land site was opened up, and the urban design requirement was for a new street frontage to heal the wounds caused by the road engineering operation. A new urban corridor of apartments over retail along Cork Street was developed with the demolition of existing social housing pockets and the need for new social housing emerged to relocate residents.

The scheme works between the six story scale proposed in general along the new Cork Street corridor and the smaller scale of the existing houses behind the site. The new buildings are in brick, with hardwood windows and screens to terraces and roof gardens. The windows are offset from each other in the walls to work with the complexity of the residential accommodation with-

N
0
10
20m

in, and to emphasize the continuity of the brick surface. The walls are modulated with recessed porches and terraces and projecting bay windows to give a depth and complexity to the building's edge and an interface between the private world of the house and the neighborhood. The building cranks along the street line with landscaped planters and steps at ground level to allow some privacy to those units accessed from the street.

The main social/play space of the scheme is the triangular courtyard which provides a secure space via the passive surveillance from the adjacent apartments. This space is further animated by the window seats at the ground level, recessed balconies and projecting winter gardens above.

The scheme opens up two new pedestrian routes through the main courtyard and the grotto at the east end of the building which knit into the surrounding urban fabric, re-making connections through the urban fabric which was extinguished by the Coombe Bypass road engineering scheme.

Brick and timber are the main materials used, echoing the existing housing and industrial buildings in the area and the former use of the site as a timber yard. The building is an in situ concrete structure with a brick skin, a typical cavity construction. The concrete structure allowed the openings in the facade to be offset from each other and also enabled a greater flexibility with apartment layouts by stepping the internal party walls vertically. The window openings recess a full brick, giving the openings as greater sense of depth. All openings have solid hanging brick lintels and brick sills to create an illusion of punching through the brick facade. The courtyard is paved in a carpet of brick to match the facade. Iroko timber windows and timber screens are left untreated.

O'Donnell+Tuomey

西南立面　south-west elevation

A-A'立面　elevation A-A'

B-B'立面　elevation B-B'

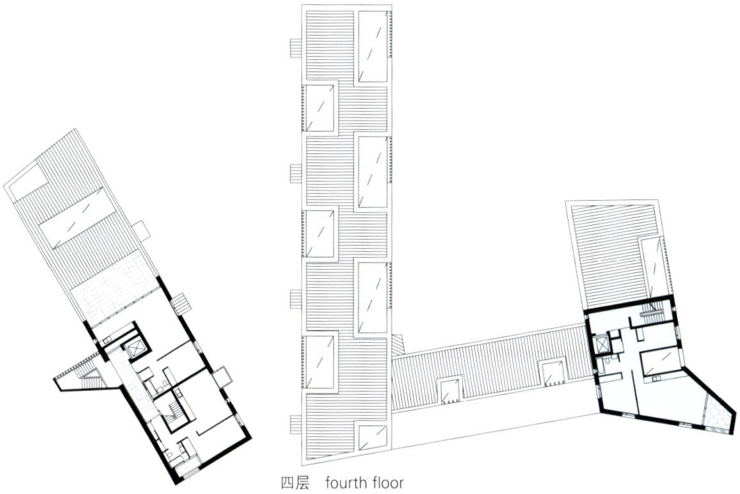

四层 fourth floor

三层 third floor

二层 second floor

一层 first floor

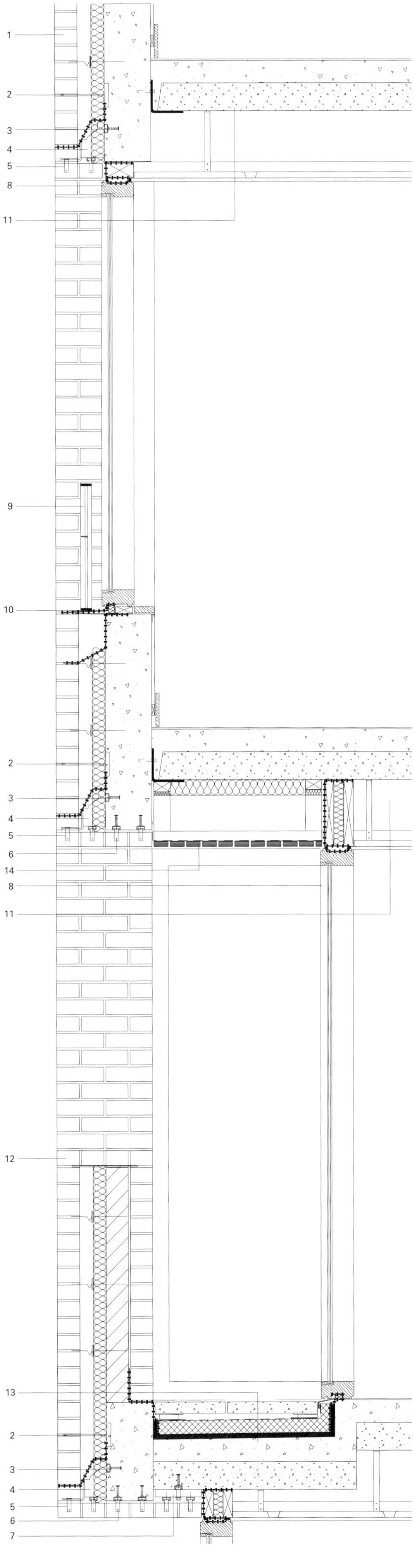


1. brick outer leaf
   lime mortar joints
   60mm kingspan thermal wall
   TW50 insulation
   215mm RC to ENG. SPEC.
   12mm sand / cement plaster
   with 3mm skin
2. halfen ribslot channel
   halfen HTS-B tie
3. 38 / 17 HTA channel
4. halfen HMA brick
   support system to
   suit loading and cavity
   with 38 / 17 HM channel
   welded to bottom
5. halfen hanger (special) c/w *6 × 70 dowel
6. 2no. 38 / 17 HTA halfen channel
7. 38 / 17 HTA channel with M12
   t bolt for hanger and HL channel
8. untreated iroko window
9. 10mm powdercoated steel balustrade
10. 4mm powdercoated mild steel
    sill with PDC dressed into
    window flashings
    on 30mm insulation
    on slate cavity closer
11. 5mm marmoleum
    100mm structural screed
    125mm precast "wideslab" floor units
    gyprock MF system suspended ceiling
    15mm plasterboard with 3mm skin
12. brick to top of balustrade
    with slate cavity closer
    and stepped DPC
    brick outer leaf
    lime mortar joints
    60mm insulation
    kingspan thermal wall TW 50
    102.5mm blockwork
    brick inner leaf
    lime mortar joints
13. 50mm precast concrete laving flag
    (450 / 450 / 50mm)
    adjustable support system
    minimum height 17mm
    minimum diameter of base 150mm
    buzon screwjack pedestial
    2mm non-woven polyester fleece
    60mm extruded polyester insulation
    thermal conductivity 0.027 W/mk
    knauf-insulation polyfoam roofboard
    20mm mastic asphalt, permaclad
    in 2 coats
    on black sheeting felt
    100mm structural screed (minimum)
    laid to falls (1: 80); max. 110.6mm
    125mm precast "wideslab" floor units
14. 70mm insulation
    kingspan thermal floor TF70 zero ODP
    44/75mm treated firings @ 300 CRS
    insect mesh
    25/75mm untreated iroko boarding
    10mm gap for ventilation

项目名称：Timberyard Social Housing
建筑师：O'Donnell + Tuomey
结构工程师：Downes Associates
估料师：Boyd Creed Sweett Limited
机电工程师：Buro Happold
考古学家：Archaeological Technology Ltd.
防火顾问：Michael Slattery + Associates
承包商：Townlink Construction Ltd.
定制外部细木工：GEM Manufacturing Company Ltd.
专业细木工：Specialist Joinery Group Ltd
扶手/栏杆：TARA Stainless
砖拱腹：Ancon System, Long Construction Services
冬季花园：Kilcoole Steel-Steelwork, Flexi Fab-Glazing
砖：Lagan Brick.Ltd
水磨石：PJ Ryan & Son
地点：Dublin, Ireland
建筑面积：3,800m²
设计时间：2001
施工时间：2007.3 ~ 2009.4
摄影师：©Dennis Gilbert/VIEW pictures

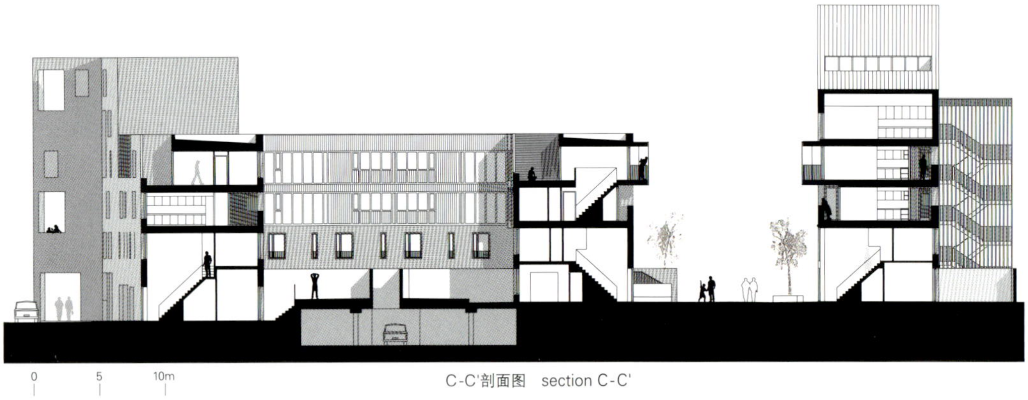


C-C'剖面图　section C-C'

D-D'剖面图
section D-D'

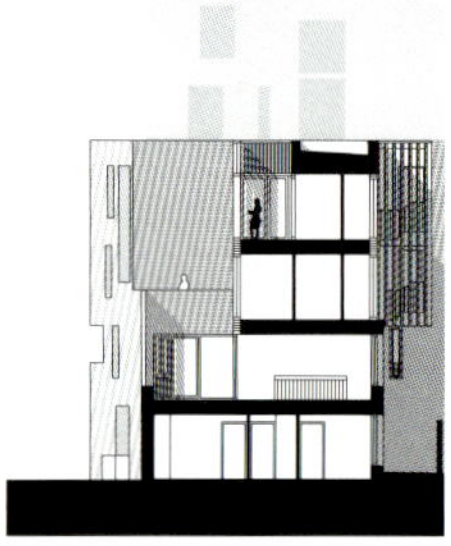

E-E'剖面图
section E-E'

# Siloetten住宅大厦

C. F. Møller Architects

丹麦很多城镇的中心都坐落着一些工业筒仓，其中大多数都不能再使用了，但还留存着，从外表上看还能构成当地的主要风景线。这种情况同样出现在奥尔胡斯市北部的Løgten小镇，在这里，原来的筒仓综合体已被转变成一个“田园式高层”，该高层包括21个高质量的住宅单元，构成了独特、别致的“叠层别墅”。

这些住宅单元是标准公寓或独立式郊区住宅的替代品，也是单层公寓和复式住宅的混合体，这种设计意味着即使较低楼层也能充分享受观景视野，也意味着任何两套公寓都不相同。

实际的筒仓结构包括楼梯和电梯，同时为公共的屋顶露台提供了基础。在塔楼四周，公寓以引人注目的形式建在一个钢结构上，公寓向外凸出到阳光能照射到的范围内和周围景观当中——这有点像乐高积木。这种带有凸出部分和移位空间的异常结构为所有的公寓都提供了广阔的户外空间，以及奥尔胡斯港和城市本身的风景。同样，无论是在筒仓结构的北面还是南面，无论是清晨、中午还是傍晚，每个公寓都能享受到日光照射。

在筒仓结构的底部，一个新的“村中心”被创造出来，该“村中心”有一个公共空间，四周环绕着包含商店、超市和联排住宅的多功能建筑体，还有一个绿色公园，公园中有分配给居民的小块土地。

筒仓结构“田园式高层”的特点是独一无二的——因为它是原有建筑的改建，该地区没有任何其他建筑能修建到与它同样的高度，所以它仍然是一座独立的里程碑式的建筑。它是一个转化多余结构的典范，既具有了新的身份，又增加了城郊的建筑密度。

在建筑面向新中心位置的一面，筒仓结构的主要部分被故意保留下来，目的是确保人们能永远记住此处的历史，也表示这种类型的结构和其他乡村历史地标，如教堂、钟楼或具有历史意义的风车等，具有同等的历史意义。

## Siloetten

Many towns in Denmark have centrally located industrial silos; most are no longer in use, but continue to visually dominate the local skyline. This is also the case in the town of Løgten north of Aarhus, where the former silo complex has been transformed into a “rural high-rise”, with 21 high-quality residences composed as individual and unique “stacked villas”.

They are an alternative to standard apartments or to detached suburban sprawl, and are a mix of single story flats and maisonettes, meaning that even the lower levels fully get to enjoy the views, and that no two flats are the same.

The actual silo contains staircases and lifts, and provides the base of a common roof terrace. Around the tower, the apartments are built up upon a steel structure in eye-catching forms which protrude out into the light and the landscape–a bit like Lego bricks.

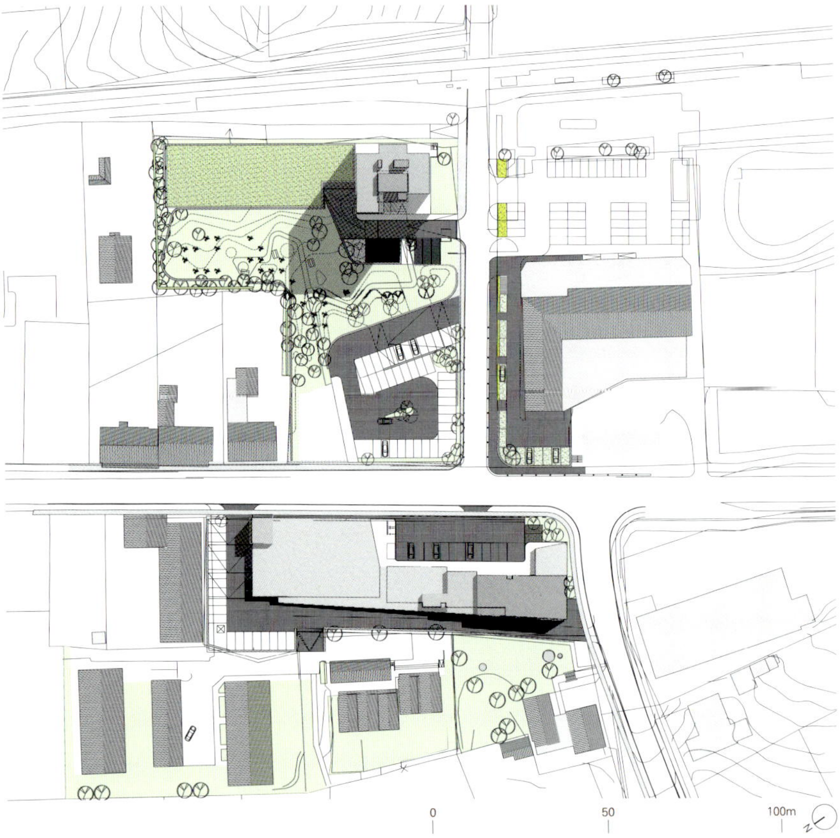
0
50
100m

This unusual structure with its protrusions and displacements provides all of the apartments with generous outdoor spaces, and views of Aarhus Bay and the city itself. Similarly, every apartment enjoys sunlight in the morning, mid-day and evening, whether placed to the north or south of the silo structure.

At the foot of the silo, a new 'village center' is created, with a public space surrounded by a mix-use complex with shops, supermarket and terraced housing, and a green park containing small allotments for the residents.

The nature of the silo's "rural high-rise" remains unique – since it is a conversion, no other building in the area can be built to the same height, and it will remain a free standing landmark. It is an example of how the transformation of redundant structures can hold the potential to both give a new identity, and introduce density to suburban outskirts.

The body of the silo is deliberately left visible on the side of the building facing the new center, to ensure a continued legibility of the history of the site, and to acknowledge that these types of structures have an equal validity as rural historical markers as do for instance the church bell-tower or historic windmills.

C. F. Møller Architects

项目名称：Siloetten / The Sil(o)houette
建筑师：C. F. Møller Architects
in collaboration with Christian Carlsen Arkitektfirma
景观建筑师：C. F. Møller Architects
工程师：Niras
甲方：Løgten Midt A/S
地点：Løgten, Denmark
基地面积：1,500$m^2$
建筑面积：silo conversion housing-3,000$m^2$
mixed-use urban center-1,500$m^2$
竣工时间：2010
摄影师：courtesy of the architect (©Julian Weyer)

五层
fifth floor

十一层
eleventh floor

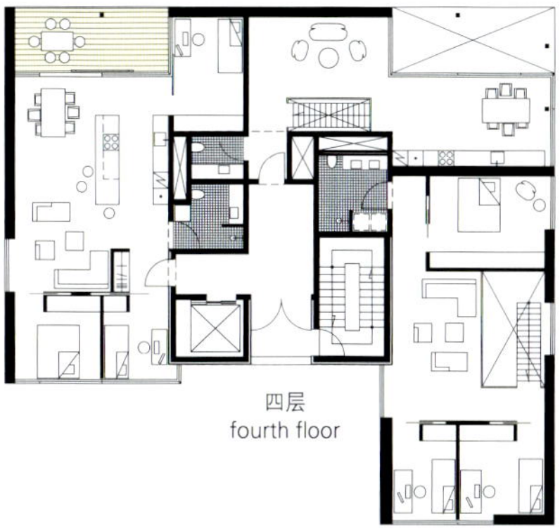
四层
fourth floor

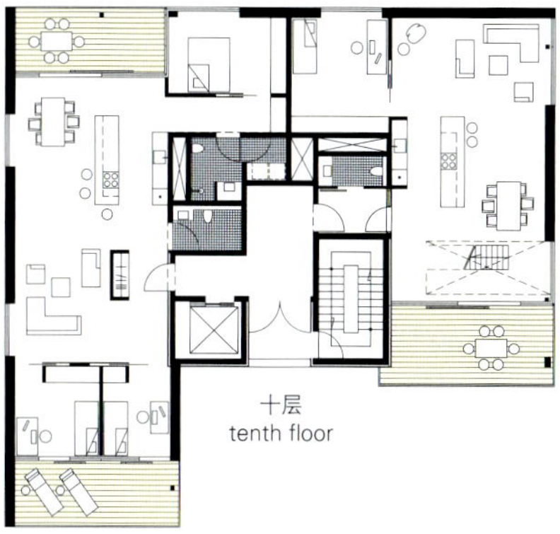
十层
tenth floor

三层
third floor

九层
ninth floor

二层
second floor

八层
eighth floor

一层
first floor

七层
seventh floor

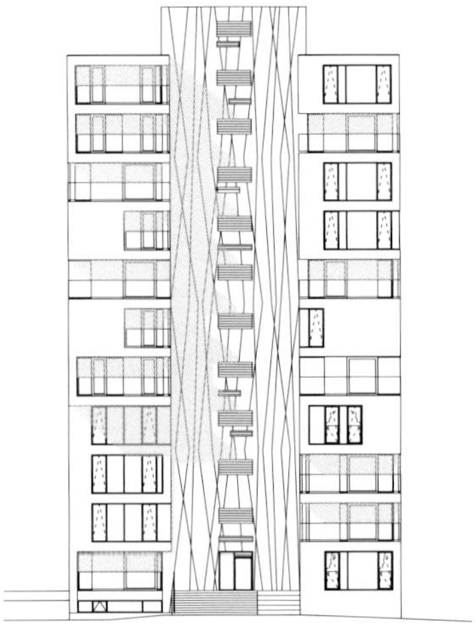

西北立面　north-west elevation

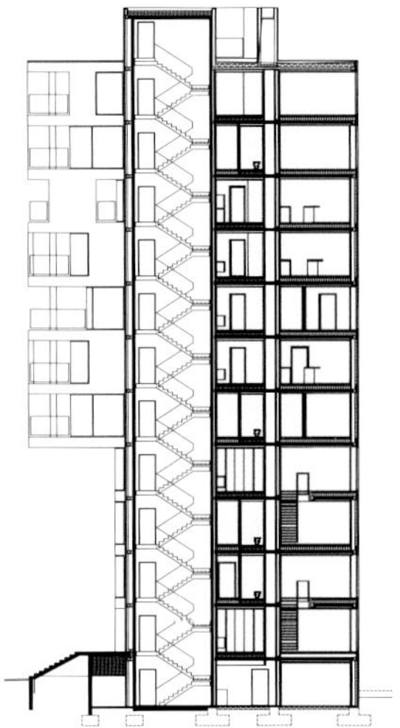

A-A'剖面图　section A-A'

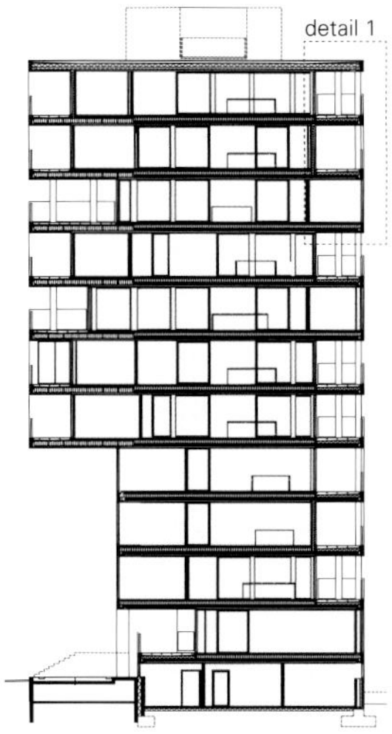


B-B'剖面图　section B-B'

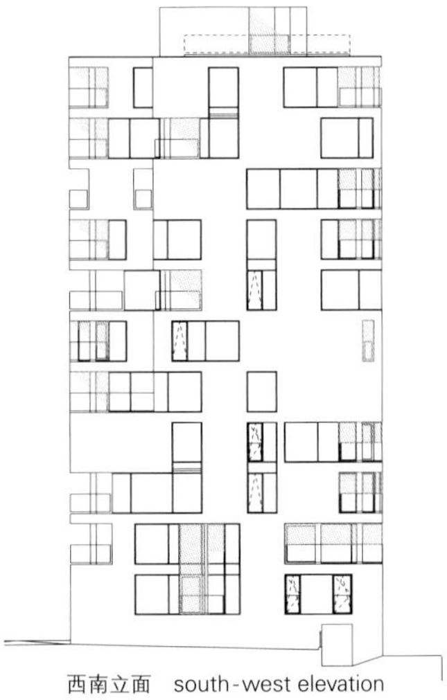

西南立面　south-west elevation

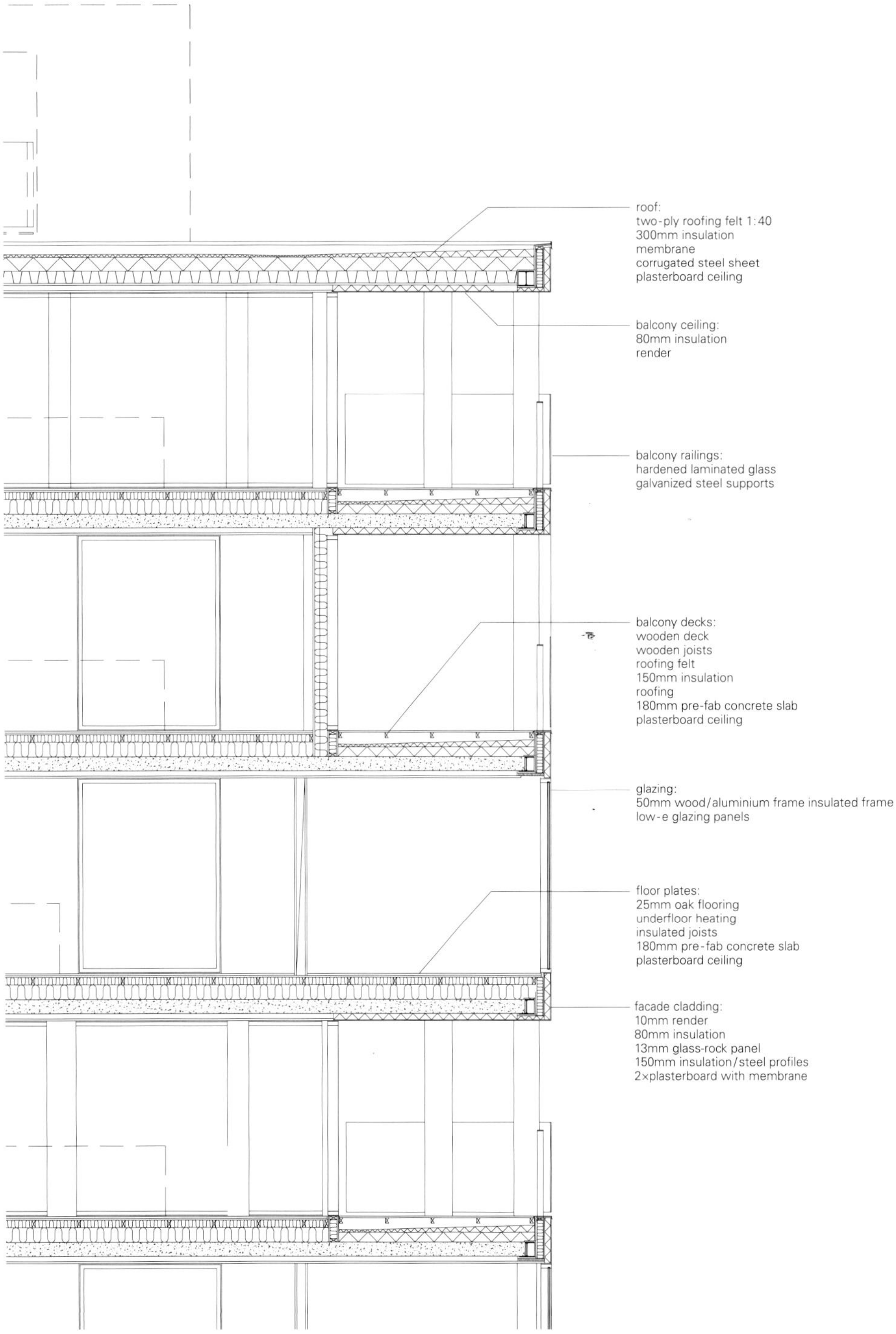


详图1　detail 1

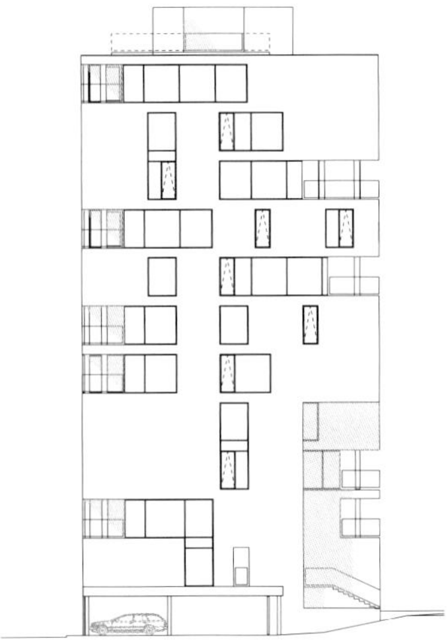

东北立面　north-east elevation

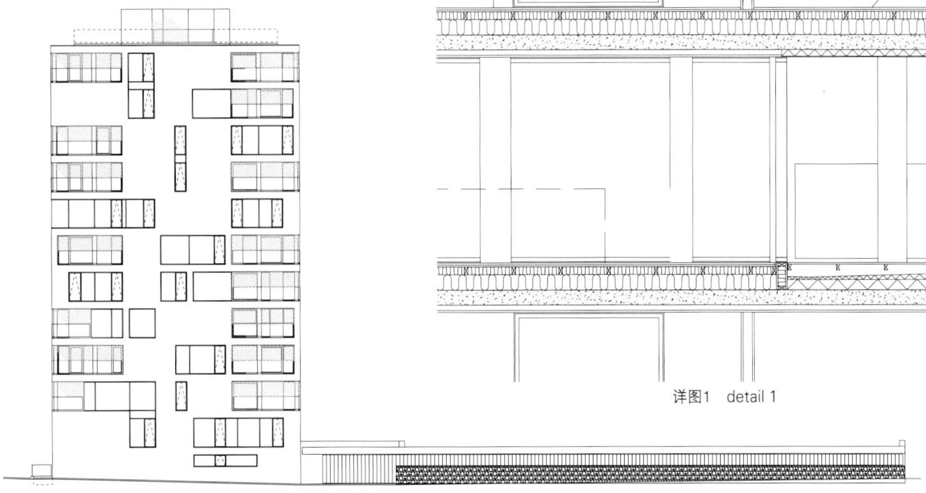

东南立面　south-east elevation

0　5　10m

# 香草园住宅

Caramel Architekten

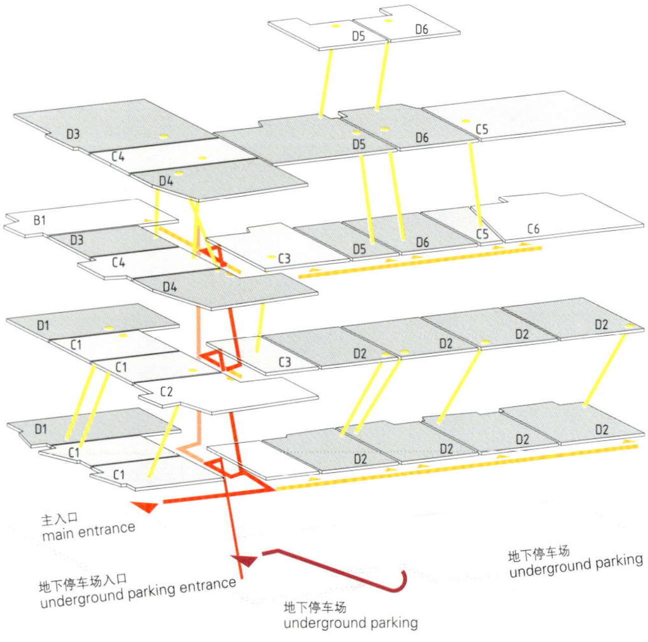


本案项目位于维也纳边缘的一个绿色区域里，四周是小型住宅和小型公寓综合体。

未来的居住者期待此处有一种“生活在绿色环境中”的氛围，理想地说就是拥有自己的房子和花园。设计的目的也是要将这种氛围传达到住宅项目本身当中。每套公寓都被设计成带有复层和不同观景视角的独立住宅。各个单元被连接起来，组合在一起，构成了整体结构。虽然krautgarten项目是一座四层的建筑，但每个居住单元都设计得尽可能独立完整。

因此，每套公寓都有自己的平面布局，各式各样的凉廊和阳台会让公寓外面的人看到每个单元都是一个独特的“房子”。私人庭院、屋顶花园、凉廊和阳台让每个公寓都可以不同程度地接触绿色环境——就像房屋四周被花草树木围绕着一样。

建筑的每个内部结构都为外壳带来一种有趣的特点，为功能性“生活”单元增添了一种积极的气氛。建筑外部使用的感觉粗糙的灰泥和明快的颜色更加深了这种气氛。沿着街道创造的一个绿色地带是公共空间和私人空间之间的缓冲区。

从城市开发的角度来说，项目位于一个极其多样的环境中。建筑的拐角处面对着公共交通区，与相邻建筑的主要方向一致。

这个位于城市边缘的带有“绿色”环境的住宅开发项目的主要目标群体是年轻家庭等。半数以上的公寓带花园，其余的则有屋顶露台或凉廊。

另外，主要由带露台的复式风格住宅构成的公寓具有高度的个性化特征，公寓之间的不同也可以从外部设计和立面上看出来。凉廊（“眼睛”）代表起居室，不同模式排列的小窗子代表卧室。

几乎位于建筑“心脏”位置、处于底层中心、邻近楼梯井的是一个多功能区域和一个带有沙箱和健身架的供儿童玩耍的区域。

公共区域有一个带小山丘的硫化橡胶玩耍区以及一个包含大型下沉区域的木质表面，下沉区域中可以玩沙子。另外，还有一个长长的既能坐也能躺的多功能区。虽然面积很小，但这个区域意欲成为居民可以会面、交谈和一起玩耍的空间。

## Herb Garden Housing

The property lies in a green zone at the periphery of Vienna and is surrounded by small houses and small apartment complexes.

The atmosphere that future tenants associate with this site is one of “living in green surroundings”, ideally: one's own house and garden. The intention is also to convey this atmosphere in the housing project itself. The individual apartments are designed as separate houses with multiple floors and various views of the surroundings. The individual units are interlocked and fit together to create the overall structure. Although the krautgarten project is a four-story building, each individual living unit was designed to be as self-contained as possible.

地下停车场 underground parking

楼梯 staircase

电梯 elevator

公寓入口 apartment entryway

内部通道 internal access

0
10
20m
N

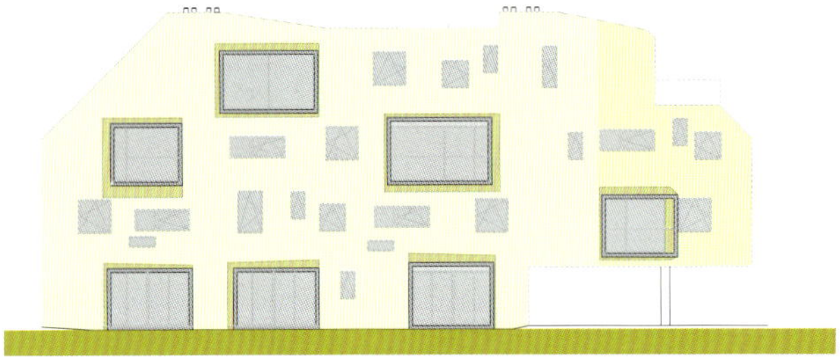

南立面 south elevation

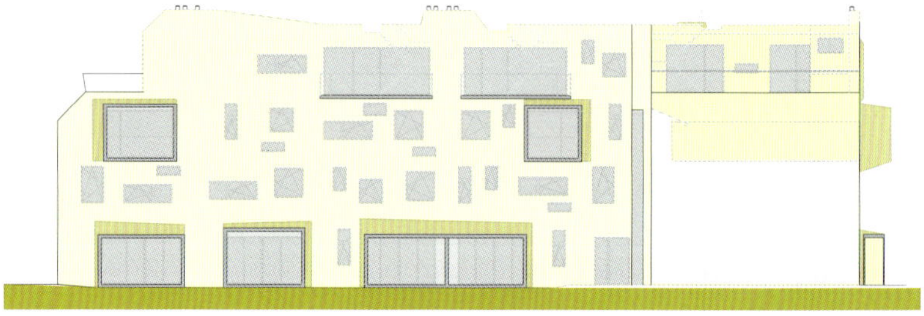

西立面 west elevation

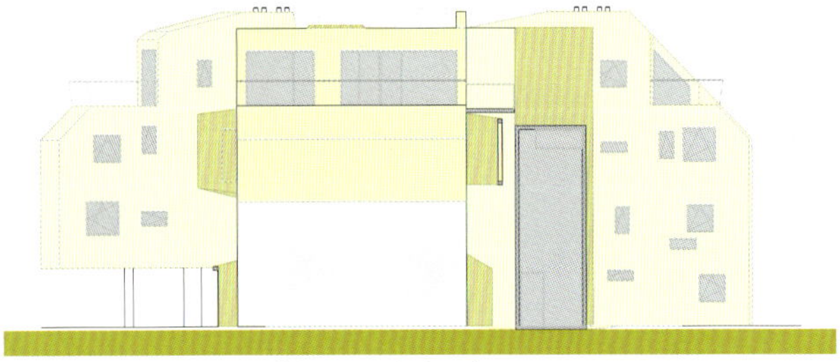

北立面 north elevation

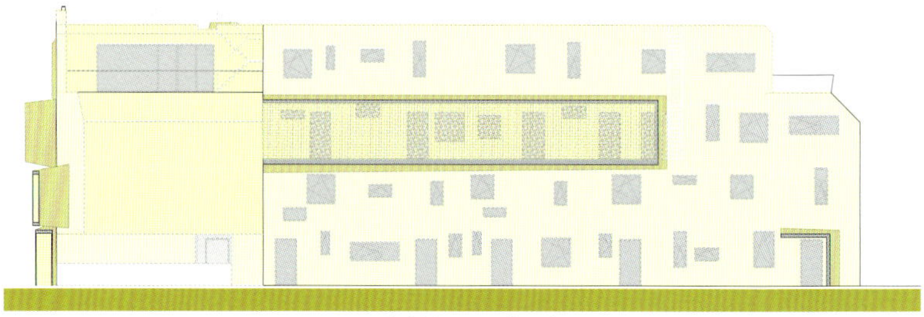

东立面 east elevation

A-A'剖面图 section A-A'

B-B'剖面图 section B-B'

0 5 10m

Each apartment, therefore, has its own floor plan and the various loggias and balconies allow the outside observer to recognize each unit as a distinct "house". Private yards, roof gardens, loggias, and balconies provide each apartment with various levels of direct contact to the green surroundings – like a house surrounded by grass and trees.

The individual internal structure of the building gives the outer shell a playful character as a positive atmosphere in a functional unit "for living". This is emphasized by the haptic impression of the very course plaster of the building exterior and the cheerful choice of colors. A green zone is produced along the street as a buffer between public and private spaces.

From an urban development perspective the site is located in a very heterogeneous environment. The corner of the building facing the public traffic area assimilates the main directions of the adjacent buildings.

The main target group for this housing development with "green" surroundings at the periphery of the city comprises young families and the like. More than half of the apartments have their own gardens; the rest have terraces or loggias.

In addition, the apartments – mainly terrace-house-style duplex units – feature a high degree of individuality; the differences in the apartments can also be seen in the exterior design and the facade. The loggias ("eyes") indicate the living rooms; the smaller windows arranged in varying patterns mark the bedrooms.

Located virtually at the "heart" of the building, at the geometrical center of the ground floor, and adjacent to the stairwell is a multipurpose zone and children's play area with a sandbox and monkey bars.

The common zone consists of a wooden surface including a large sunken area for playing with sand and a vulcanized rubber play area with a hill. In addition, there is a long multipurpose element good for both sitting and lying on. Despite its small size, the area is intended as a space where residents can meet, talk, and play together. Caramel Architekten

一层 first floor

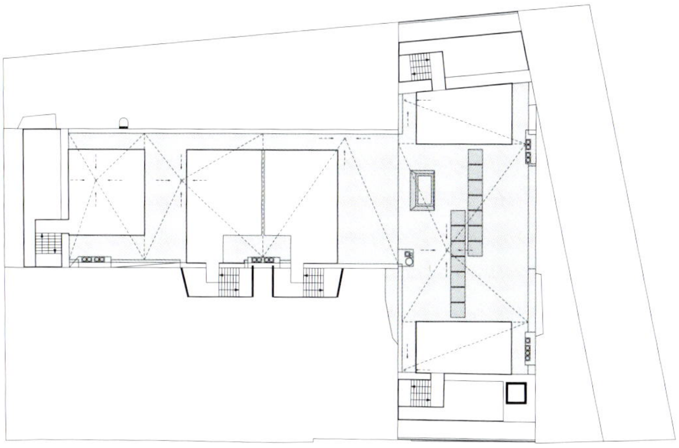
屋顶 roof

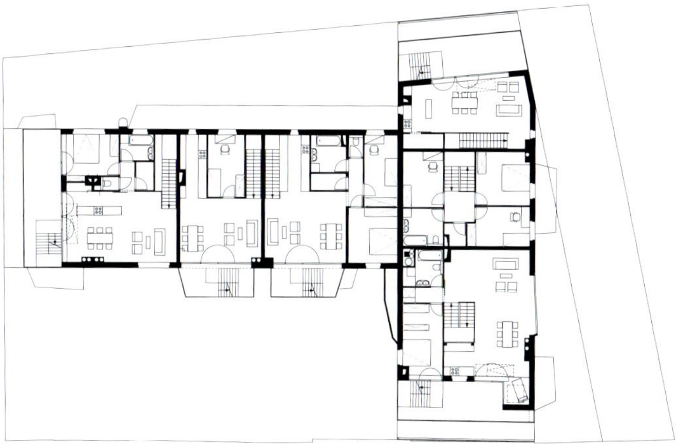
四层 fourth floor

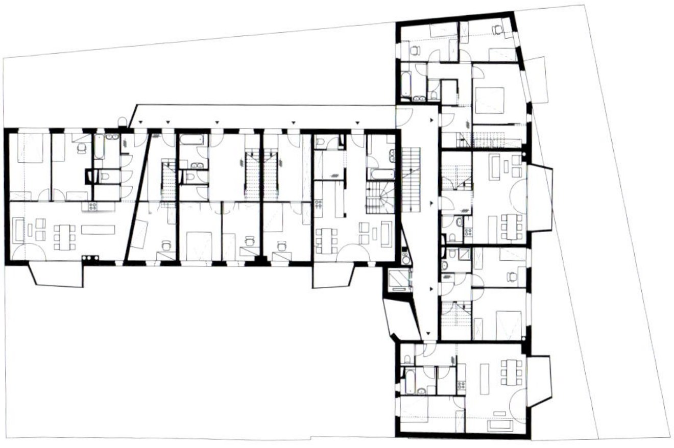
三层 third floor

二层 second floor

项目名称：Housing Development at the Herb Garden
建筑师：Caramel Architekten
景观建筑师：Dealice-Alice größinger
总体规划：Stumpf Develepment Gmbh
甲方：Bauhilfe Gemeinnützige Gmbh
地点：Vienna, Austria
总建筑面积：2,705m²
设计时间：2007.12
施工时间：2009.3~2010.12
摄影师：courtesy of the architect (©Hertha Hurnaus)

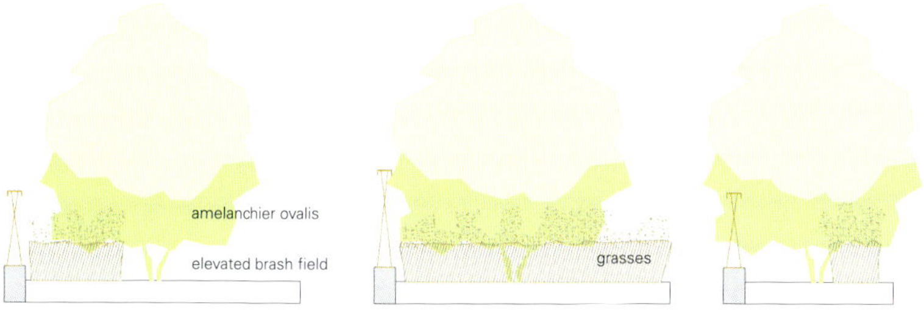


栅栏构件和延伸到街面的绿色缓冲区剖面图
section fence element and green buffer range to street

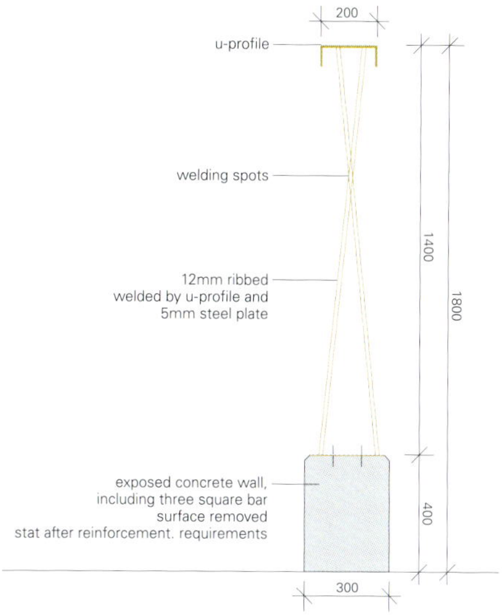


栅栏构件设计详图
detail design fence element

面向街道的栅栏构件　fence element face to street



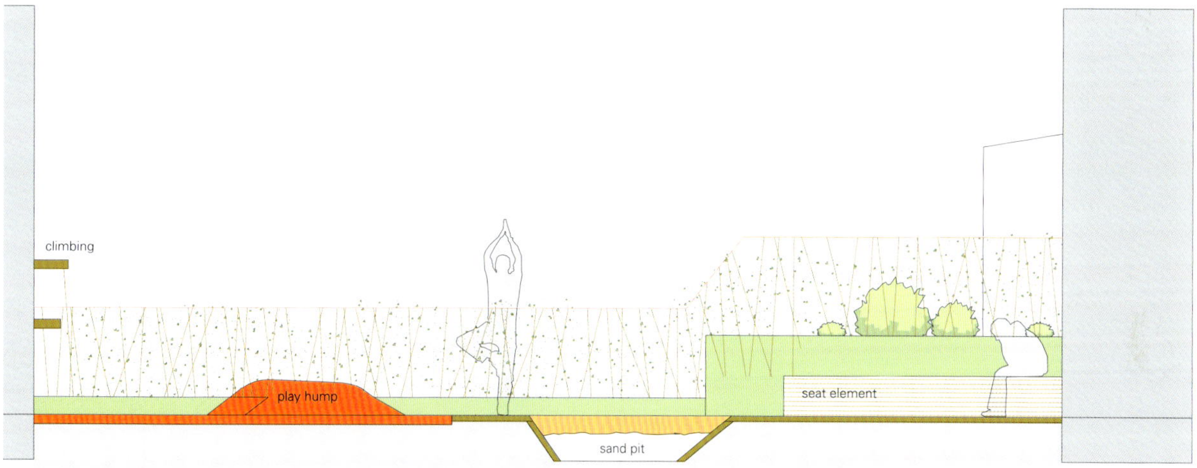


运动场与私人花园的分隔构件　partition playground – private garden

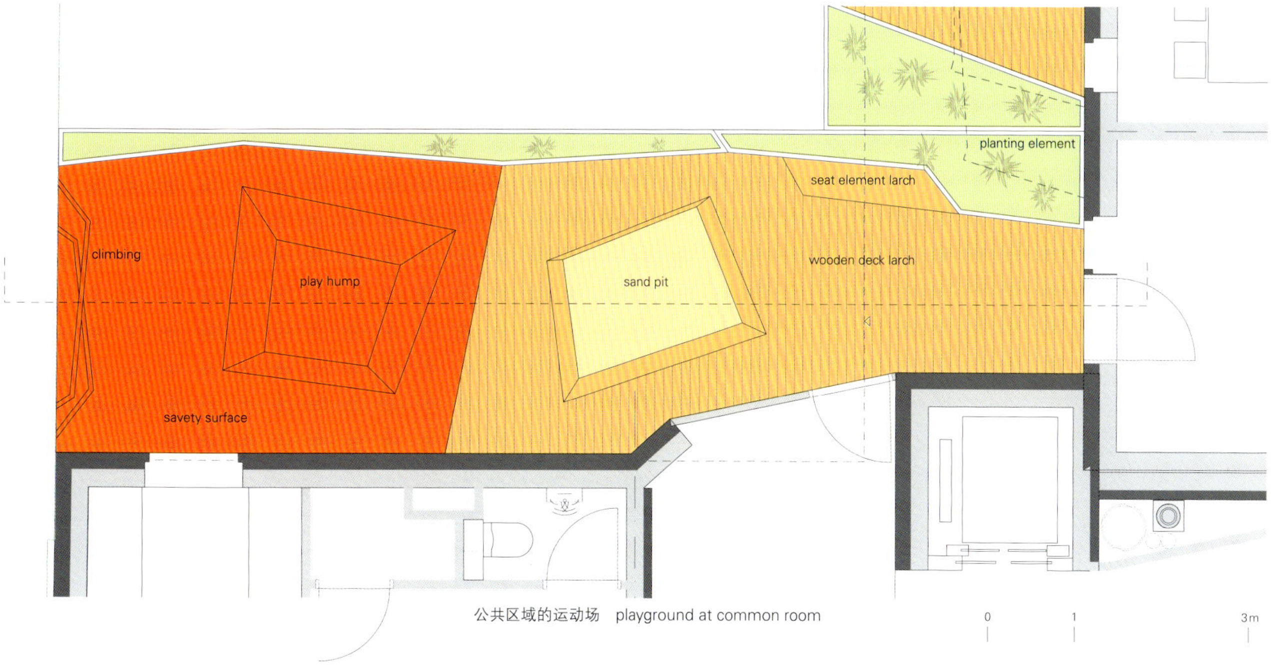


公共区域的运动场　playground at common room

# BIG 未来之窗

# BIG Window to the Future

采访者Michele Stramezzi

Michele Stramezzi（下文简称为：M.S.）：我第一次接触BIG（Bjarke Ingels Group）的作品是在五六年前，当时我正在为一个新的住宅项目找些参考案例，我找到了"VM住宅"——他们第一批建成的建筑之一。

为住宅设计有趣的立面很具挑战性，因为表皮后面的功能空间通常展示出单一的开窗规律。如果业主还要加上阳台，那么你就有麻烦了，你可能只是试图把阳台设计得大方而忘记了它本身应具有的吸引力。阳台是时髦立面的噩梦，很少有设计师能做出什么花样。

"VM住宅"看上去并不高档，它建在高速路边上，而且阳台就像是要被飞驰而过的汽车带过的的风刮掉了一样。这样的效果简单而又新奇，小阳台蕴含的无尽创造力就像是对美好未来的前瞻。

BIG：经济危机前的那几年，全球房地产市场很繁荣，当时我们大部分的设计都是服务于私人开发商的。他们允许我们在住宅、酒店和办公等普通项目这一相对狭小的领域内发掘建筑的潜力。

我们有幸在哥本哈根做项目——例如"山阶住宅"和"8字之家"，在这些项目中我们能够对普通项目进行不同的创新尝试，并试图彻底改造传统复合功能街区。同时，飞涨的价格造成的不利影响使城市住房价格越发不能为普通收入家庭所接受。所以我们尝试前瞻性地提出对城市内部地区进行密集化设计的新形式，设计了一系列更具挑战性的项目，如"乐高塔"、"三叶草地"和"电池组"，这些项目得到了国际性的关注。

从某种意义上来说，我们是试图在开发商的利润和丹麦集体福利价值之间穿行。与障碍物不同，这成为设计新形式的有效推动力。

经济危机后，我们看到多年的狂热开发向关注城市面临的问题的转变。很明显，对于私人投资和公共投资来说，生态和经济是一个硬币的两面，相互依存。繁荣时代带来乐观和对美好的信心，这无疑是一个必要的、健康的转变。

M.S.：BIG可能是世界上最年轻的优秀建筑公司，他们对建筑和

**Michele Stramezzi** My first encounter with the work of BIG – Bjarke Ingels Group happened five or six years ago: while researching for references for a new residential building project, I found some images of the VM House, one of their first realized architectures.

It is a challenge to produce something particularly interesting with the facade of a housing building: the functional spaces behind the skin usually dictate a rather dull windows opening rhythm. When your client is asking for balconies, well, then you are in trouble; just try to make it decent and forget about glamor. Balconies are the ruin of the Funky Facade; very few designers can make something nice out of them.

The VM House did not or does not look like an expensive building; it is built close to the highway and the balconies seem like driven away by the wind produced by the fast cars driving by. The result is simple and surprising at the same time, the amount of creativity inside that little balcony trick was foreseen for a bright future.

**BIG** In the pre-financial crisis years the global real-estate market was booming, and we were doing mostly work for private investors, allowing us to explore the architectural potentials within a relatively narrow field of common programs like housing, hotels and offices.

We were fortunate to have construction work in Copenhagen – in projects like the Mountain and the 8 House we were able to test ways of composing ordinary programs, and try to reinvent the traditional mixed used urban blocks. At the same time the sky-rocketing prices had the negative side effect, that housing in the city was becoming increasingly inaccessible for normal-income families. So we were experimenting with proactively proposing new forms of densification of the inner city areas, resulting in a series of more ambitious projects like Lego Towers, the Clover Block or The Battery that gathered international attention.

In a sense we were trying to navigate between profit hungry investors and collective Danish welfare values. Rather than an obstacle this turned out to be an effective fuel for inventing new typologies.

After the crisis we saw a change from years of development frenzy to a new focus on solving the problems that cities are facing. Both in terms of private and public investments it seemed to be clear that ecology and economy are two sides of the same coin. While the booming years did lead to optimism and a new belief in the fantastic, this was obviously a necessary and healthy change.

**M.S.** Being probably the youngest amongst world's most prominent architecture firms, BIG own point of view about the relationship between architecture and society, especially when related to a projection and understanding towards the future, is as intriguing to understand as much as their creativity in the approach to

社会的关系，尤其是涉及对未来的理解上有自己的观点，这些观点连同他们在建筑设计中的创造力都很耐人寻味。

BIG：在一个始于2004年的研究项目中，考虑到所有的现实约束条件，我们用名为"务实乌托邦"的词来说明我们重新组织或重新设计社会以大幅提高生活质量的信心。这一想法能够更全面地诠释我们的理想。

公司的发源地在丹麦，因此我们注重务实和对环境的格外尊重。在设计和沟通中，对集体和公共利益的注重始终贯穿我们的设计。

M.S.：建筑师最能通过设计表达他们对社会的结构和发展观点的领域当然是大规模的总体规划。

"环形城市"是BIG在上届威尼斯双年展上对城市区域尺度发展所作的研究。针对丹麦和瑞典之间的斯堪的纳维纳地区，这个规划从不同尺度上体现了公司对可持续发展以及建筑、使用者和所属区域关系的认识。"环形城市"和罗马输水道形式相似，并让人回想起缩小版的勒·柯布西耶的"阿尔及尔城市规划（1933）"方案。与线性城市沿基础设施沿线发展类似，这里是将顺序倒置，城市成为建筑师用以确定新基础设施线路的点。

BIG：我们出版了针对哥本哈根区域发展的研究：环形城市。哥本哈根50年来一直遵循手掌形的图示发展，五根手指指示了首都的发展方向。在"环形城市"项目中，我们提出了一种新的瑞典和丹麦间跨区域发展模式——这个环不仅是人类的基础，而且是能源、废弃物、水和生物多样性的基础。"环形城市"将成为这个图示很好的示范——它不仅在外形上，而且在内在逻辑上都是可识别的。从这个意义上来讲，它没有抛弃图示，而是关注外形而非图像。

M.S.：为了2010年上海世博会开幕式，我特地见了比亚克·因格尔斯（Bjarke Ingels）。在上海，BIG公司正在展示他们做的丹麦国家馆方案，并要带所有员工到中国共同庆祝这一盛事。在我和他们的老

the building architecture design.

**BIG** In a research project from 2004 we used a term Pragmatic utopias – the ambition to reorganize or redesign society in order to drastically improve the quality of life – while taking all practical constraints into account. This idea could be describing our ambition in a broader sense.

Being based in Denmark, we have been used to operating in pragmatic and strongly consensus driven environment. The importance of addressing the collective and public interests, both in terms of design and communication has been fueling our projects.

**M.S.** The field in which the architect can express through the tools of designing their idea about the structure and development of the society is of course that of the big scale masterplan.

Loop City is a study on regional scale urban development which BIG presented at the last Venice Biennale. Applied on the Scandinavian region between Denmark and Sweden, the plan shows from different scales the position of the Firm about sustainable development and relationship between architecture, its users and the territory on which it is inserted. Loop city has quite the shape of a Roman aqueduct and reminds of Le Corbusier scheme of "Urban Plan for Algiers(1933)" reduced in scale. Like the Linear City is developed along the infrastructure line, here, the other way around, the cities are the points which the pen of the architect connects with the new infrastructure line.

**BIG** We have published a study on the regional development of Copenhagen: Loop City. The city has for 50 years been developed according to an icon: The Finger Plan, the image of a hand with five fingers guiding the development of the capital. In the project Loop City we are proposing a new growth model as a cross-border region between Sweden and Denmark – A loop of not only infrastructure for people, but for energy, waste, water and biodiversity. The Loop City could be an example of the power of the icon – it is recognizable, not for its looks but for its inherent logic. In that sense it is not a question of abandoning icons, but to focus on the performance rather than the image.

**M.S.** I met in person Bjarke Ingels and his Group of Architects for the opening of the 2010 Shanghai Expo. In Shanghai the Firm was presenting their design for the Danish National Pavilion and decided to take all the employees over to China to celebrate this event. While I was talking to the boss, the rest of the team were either enjoying a drink on the roof of the building or biking up and down the ramps crossing the pavilion. It was in this occasion when I heard for the first time Bjarke Ingels talking about a concept which later again will appear in the presentation of other of their designs: Hedonistic Sustainability.

*"The idea is that all the discussion about Sustainability doesn't neces-*

板谈话时，团队的其他人或者在屋顶平台上享受美酒，或者骑着自行车沿展馆的坡道上上下下。就在这种场景下，我第一次听比亚克·因格尔斯谈起“快乐可持续”，这一概念也将在他们其他的设计中出现。

“这一概念就是：所有关乎可持续的讨论都和投资绿色环保所要付出的代价的多少无关。我们要改变这一情况，如果可持续只是人们出于道义责任才能达到的目标，那么它就失去了吸引力，我们试图去展示可持续的城市能够真正成为提高生活质量的途径。”

“快乐可持续”的想法还体现在哥本哈根市郊废物能源电厂的设计中。在这里，人类为了处理废品不得已而为之的情形被转化成娱乐项目：市郊人工滑雪场。在建筑作品和娱乐项目之间，发电厂燃烧废品排出废气时也自得其乐——废气从烟囱排出时形成环状的烟圈，就像爷爷叼着烟斗或绅士享受名贵香烟时吐出的那样。彩色的激光束打到空中的烟圈上形成悬浮的光之雕塑。

BIG：工厂主管希望这座建筑能吸引公众的注意，并与港口的娱乐性景观相结合。我们计算了建筑的高度足够做成一个约500m高的滑雪坡道。所以这个废品厂不仅为哥本哈根提供了二氧化碳中性气体，也成为丹麦第一座滑雪山。

对于我们来说，这个项目可以作为解释“快乐可持续”的最终符号。可持续概念不只是减少资源的利用、放弃现代的生活方式，还应该真正地提高生活品质。

M.S.：BIG工作室赢得竞赛的比率几乎无人能及。不久前BIG赢得了另一个竞赛：格陵兰岛新国家美术馆。他们将一个简单的几何图形置入优美的环境中。环形建筑三维地随海湾波浪起伏的岩石地形而变化，外部封闭，内部敞开，陡峭地形所造成的不同高度使它可以俯瞰下面的美景。

如同世界上大多数建筑师一样，BIG更愿向西方拓展而非东方。最新消息说BIG将在纽约开设工作室，地点就在西57街——他们即将开工的有600个单元的住宅大楼旁边。这座大楼采用了基本的几何形

*sarily have to be just about the welfare you have to give up in order of financing a green approach: how much welfare are we willing to give up to be green. We need to flip this the other way around; if Sustainability is only something you achieve out of like moral obligation it becomes unattractive; we try to show that a sustainable city can actually be a way of increasing the quality of life."*

The idea of Hedonistic Sustainability is again expressed within the presentation of the Waste to Energy Plant outside Copenhagen. In this situation, what is usually the necessary and unpleasant addiction to a human conglomerate in order to destroy its own waste is transformed into an amusement item: a semi urban artificial ski resort. Standing between the work of architecture and the pure divertissement, the Energy Plant plays with its own waste from waste: the exhaust gas leaves the chimney shaped as smoke rings, like would do from the pipe of grandpa or from the mouth of a gentleman playing with the smoke of his expensive cigar. Colored laser beams reach the rings out in the air producing floating light sculptures.

**BIG** The director of the plant wanted to create a public attraction, and to integrate the facility in the recreational landscape of the harbor. We figured the building was tall enough to form a skiing slope of almost 500 meters. So the waste plant could not only supply Copenhagen with $CO_2$ neutral energy, but also become the first skiing mountain in Denmark. For us this project has been the ultimate symbol of what we have been calling hedonistic sustainability. The idea that sustainability is not only about reducing use of resources, and resorting to a less modern lifestyle, but should actually be about improving the quality of life.

**M.S.** BIG is the kind of office who wins competition at a rate which probably nobody else does. Another competition BIG has won in the past few months is the Greenland's New National Gallery, a simple geometrical figure inserted into a beautiful powerful natural context. A circular building adapted with its third dimension to the undulated rocky fjord morphology; closed to the outside and transparent on the inside, overlooking its beautiful scenery thanks to the difference in the height produced by the steep site.

The Firm moves west instead of going east as almost any other architect in the world is doing. Latest news about BIG is they are opening an office in New York City, location where they will also start building a 600-Unit Residential Building soon on the West 57th street. Basic in its geometrical definition, the building, rectangular on the floor plan and triangular in the elevation, combines the typical Manhattan high rise typology with the European perimeter block community.

**BIG** The US has traditionally been dominated by on one hand generic and practical developments, on the other hand visually sophisticated and expensive architecture. In this environment we be-

状，有长方形的平面和三角形的立面，结合了典型的曼哈顿高层建筑原型和欧洲围合建筑社区。

BIG：传统上美国被两种方向所支配：一方面是普遍而实用的开发项目，另一方面是外表复杂而昂贵的建筑。在这种环境下，我们相信将实用和有效相叠加能提供一种新的方式。

现在我们在亚洲、欧洲和美国都有建设项目。我们最初在美国设工作室是因为正在做曼哈顿西57街的住宅塔楼项目。我们希望在美国有更多机会。这里有良好的创作环境，而且后金融危机似乎将对真实的城市挑战做出新的回应。另外在亚洲，我们看到了从利益驱使和快速建设向追求高水准建筑的转变。也许2008年奥运会是这一转变的拐点——在这期间，中国建成了一些全球性的重要建筑。这一转变也在我们所挑战的项目中体现出来。今年夏天，我们开始了深圳能源大厦的建设，这个项目对建筑和可持续都有很高的要求。我们希望继续在亚洲工作，参与到城市持续扩张的挑战中。

后记：

BIG的作品总会展示有序的分析图来说明建筑背后的理念和观点：形体与环境的关系，原型界定，内外关系定义，语境诗意的互动。建筑的概念化方式和图表的表达始终保持着并随项目不断变化。对同一概念的解释有时用小图示比其他方式更有效。

BIG处理建筑问题简练而又精锐：几何、关系和功能。创作中这种对设计原则的精锐尤其必要，就如奥卡姆剃刀原理（Occam's Razor）："如无必要，勿增实体"（Plurality should not be posited without necessity）。没有多余——这是解释无限创造力和适度精简在BIG的作品中共存的唯一方式。

lieve that the overlap of practical and impactful offers a new way. We currently have construction work both in Asia, Europe and the US. In New York City we are working on the West 57th housing tower on Manhattan, which was the initial reason to establish the office there. We are experiencing a lot of new opportunities in the US. It is a great climate for invention, and post-financial crisis there seems to be a new-found ambition to address real urban challenges.

In Asia on the other hand, we have seen a shift from profit driven and fast paced building activity, to a much higher ambition for architecture. Maybe the Olympics in 2008 were the turning point for this change – some of the most important new buildings globally were built in China these years. This change is also reflected in the kind of assignments we are challenged with. This summer we started the construction of the Shenzhen Energy Building, a project with a high ambition for both architecture and sustainability. We expect to be working continuously in Asia, taking part in the challenges of the still explosive urban growth.

## Epilogue

BIG's work is always presented with a basic sequence of pictures illustrating the ideas and concepts behind the building realization: volume contextualization, typology definition, internal and external relation systems definition, objecting à reaction poetique. The conceptualized approach to architecture and the possibility to express it through diagrams and schemes has been around since a while and with alternate fortunes. Sometimes that little finger pushing the slab out to explain to us again the same idea seems more a pretext to explain the research of an iconic effect through a banal use of expressive tools than anything else.

There is a certain sharpness in the simplicity with which BIG approaches the constituting matter of architecture: geometry, relations and function. The sharpness inside a principle is extremely necessary when producing a piece of creativity, the Occam's razor: "Pluralitas non est ponenda sine necessitate (Plurality should not be posited without necessity)". This is the only way to explain how they can coexist in BIG architecture production the impressive amount of creativity and the controlled simplicity: the absence of unnecessary.

Michele Stramezzi：2003年于米兰理工建筑与城市专业获硕士学位。曾于1997年到代尔夫特理工大学做访问学生。在荷兰，曾在de Architekten Cie建筑师事务所、埃里克·范·埃格拉特联合建筑师事务所（Erick van Egeraat associated architects）和MVRDV任项目主管。移居中国北京后，他从事自由职业、私人建筑师和建筑顾问的工作，并且一直在维斯马专业大学和代尔夫特理工大学辅导硕士生。

**Michele Stramezzi** received master's degree in Architecture and Urbanism at Politecnico di Milano in 2003. Was a visiting student in the Faculty of Architecture, TU Delft in 1997. Worked in The Netherlands, amongst others, de Architekten Cie, Erick van Egeraat associated architects and MVRDV as project leader. After moved to Beijing, China, he is working currently as freelance and private architect and architectural consultant. Has been mentoring master's degree students at Technische Hochschule Wismar and TU Delft.

# Stockholmsporten规划

BIG + Grontmij + Spacescape

斯德哥尔摩北部E18和E4两条高速路立体交汇处正在规划"Hjulsta"交通枢纽。高速路对周边住宅区形成自然和视觉上的阻碍，并且把这一区域分成了四块。斯德哥尔摩Stockholmsporten竞赛试图通过塑造周边景观以及构架交叉口的交通尺度来塑造"Hjulsta"枢纽。

"斯德哥尔摩球"通过一个景观环连接了这四个部分。一个连续的自行车和步行环路连接了公共建筑和设施，以平等、民主的方式将不同的地区重新连接起来。

环内引入了不同特色的自然景观，中央的"能量谷"形成了由松树和橡树森林、湿地、草地和小山丘组成的扇形公园，为走进或沿景观区行走的人创造出不同的活动体验。周边居住区仍有扩展的空间，因此扩大的基础设施和开发项目形成了"能量谷"的山脊。

"能量谷"集成了城市、景观、建筑、艺术和基础设施，形成了斯德哥尔摩的新住宅区。利用大规模隧道和高速路的开发，将过度的挖掘转化为一条人工峡谷，我们创造了一个合乎逻辑、具有经济效益、环保的社会基础设施综合体。新的立面和建筑与Hjulsta现有的建筑结构融为一体。(在公寓配比上)着重于在现有的中型建筑基础上增加小型和大型的公寓。

项目所在地很适合作为斯德哥尔摩的门户。这里被转变为一个反光、自悬浮的球体，真实地反射出新旧共存的斯德哥尔摩景观，为进出城市的司机提供了180°的景观视角。30%的球面覆有朝向太阳的光电板，其电量除能够支持自身悬浮外还能为住宅区的235栋住宅提供电能。"斯德哥尔摩球"是标示城市入口的转换符号，反射着不断变换的季节和它下方不断改善的城市生活。

## The Stockholmsporten Master Plan

The traffic junction "Hjulsta" is being planned in the north of Stockholm, where the two highways E18 and E4 Stockholm By-passes meet in a three level intersection. The roads create physical and visual barriers between the surrounding neighborhoods and divide the area into four parts. The Stockholmsporten competition seeks to define the Hjulsta intersection through sculpting the surrounding landscape and framing the automotive scale of the intersection.

The Stockholm Sphere connects the four parts with a landscape loop. A continuous circular bike and pedestrian path aligned with public buildings and functions reconnects the different areas in an un-hierarchical and democratic way.

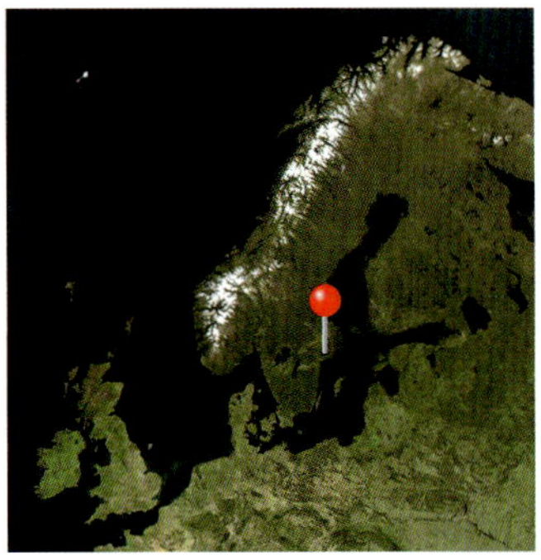

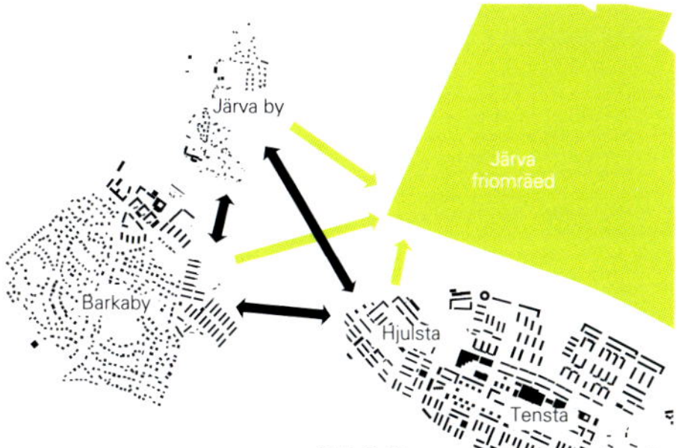

最佳连接
preferred links

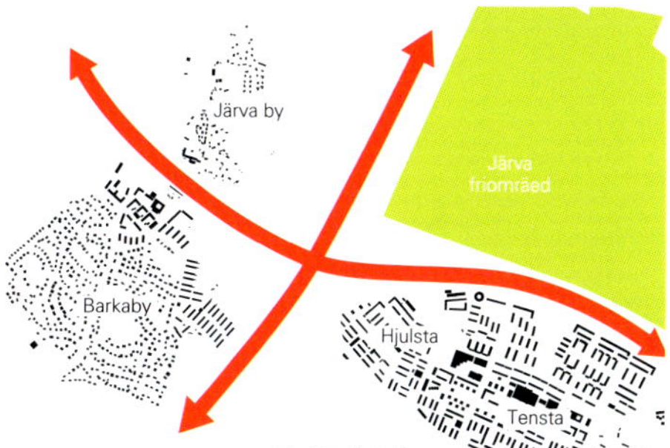

基础设施关卡
infrastructural barrier

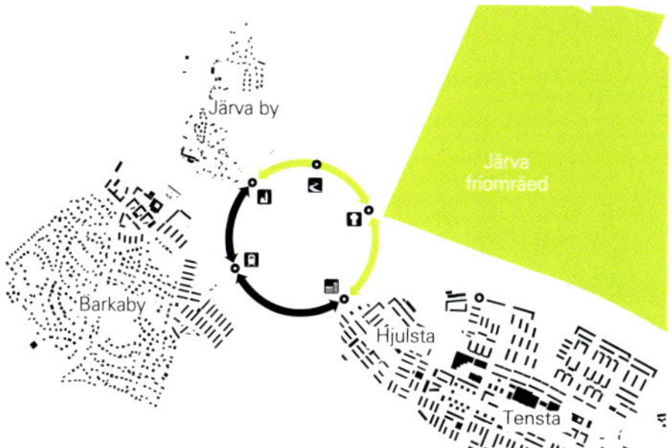

绿环
green loop

By introducing natural environments of differing characters within the loop, an Energy Valley in the center turns into a pie shaped park of pine and oak forests, wetlands, grass lawns and hilly terrain which create a diverse experience when moving in or around the landscape. The surrounding neighborhoods have room to grow, thus expanding infrastructure and developments up to the ridge of the new valley.

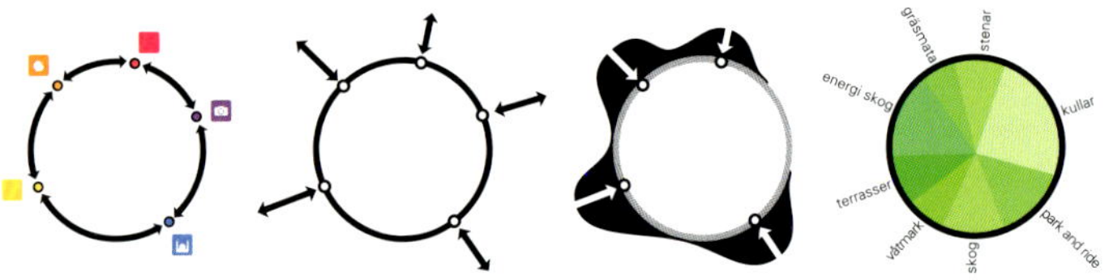


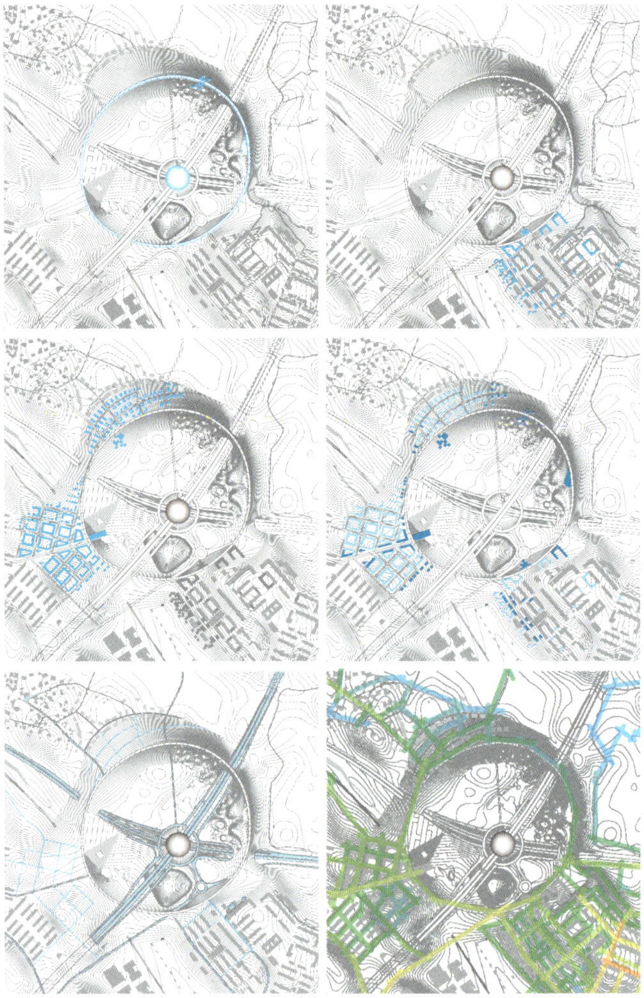

The Energy Valley is a cross-over between urbanism, landscape, architecture, art and infrastructure into a new neighborhood of Stockholm. Harnessing the momentum of the massive investment in tunnels and highways and putting the excess excavation to use as a man-made valley, we create an interdisciplinary hybrid of logistic, economic, environmental and social infrastructure. New facades and buildings are integrated in the current building structure in Hjulsta. Focus is on adding small and large apartments to the existing midsize range.

The site has a great potential in serving as a new entryway into Stockholm. This point is turned into a reflective, self-sustaining hovering sphere mirroring Stockholm as it is, new and old, creating a 180 degree view of the area for the drivers on their way in or out of the city. 30% of the sphere's surface is covered with Photovoltaic film that faces the sun and produces enough energy to keep it floating while supplying 235 houses in the neighborhood with electricity. The Stockholm Sphere is an ever changing icon that marks an entry point to the city and reflects the passing seasons and the evolving urban life beneath it. BIG-Bjarke Ingels Group

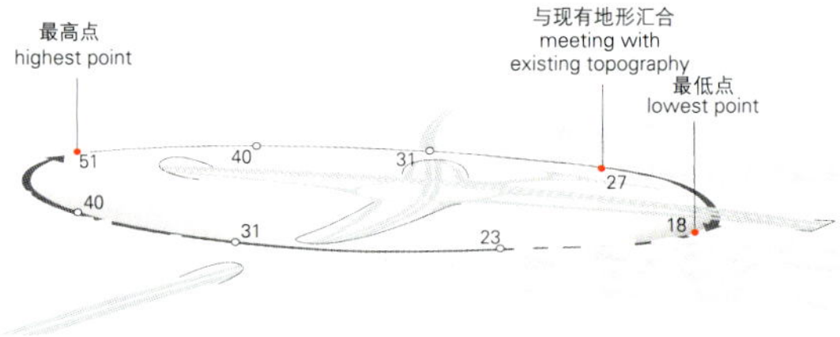


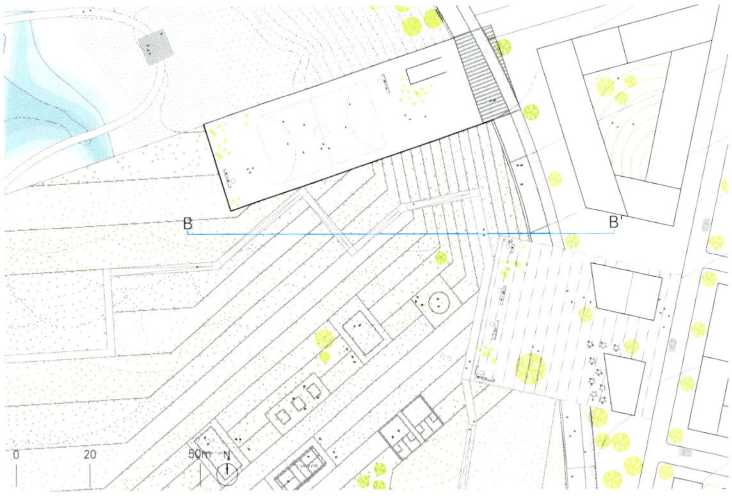


体育中心 sports center

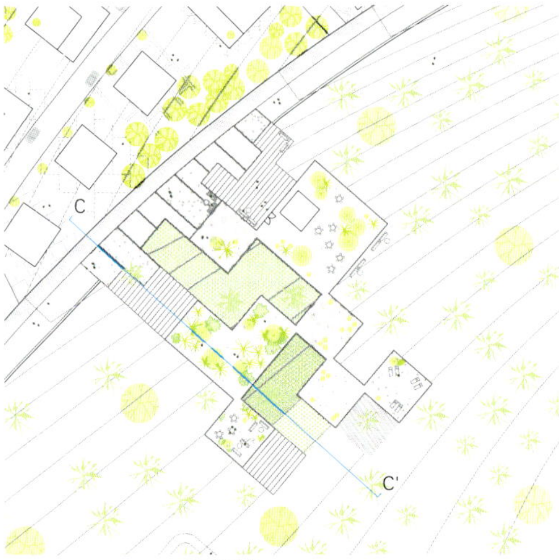


玻璃暖房 conservatory

B-B'剖面图 section B-B'

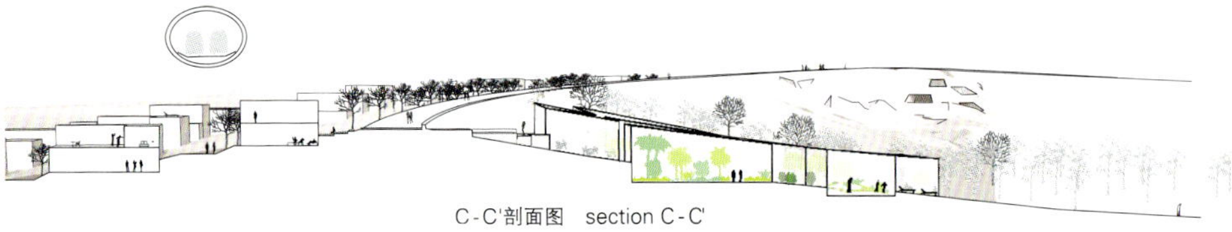

C-C'剖面图 section C-C'

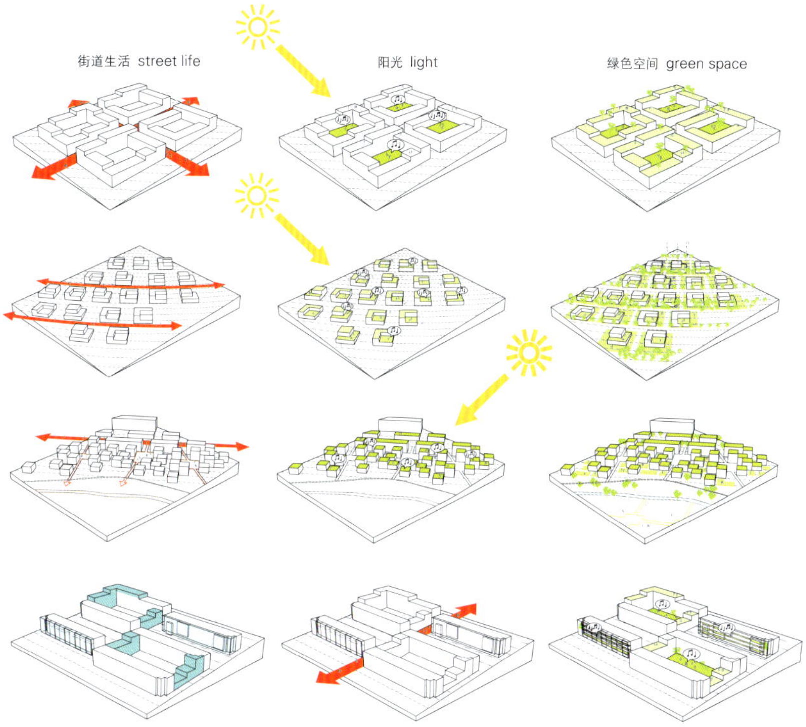
街道生活 street life
阳光 light
绿色空间 green space

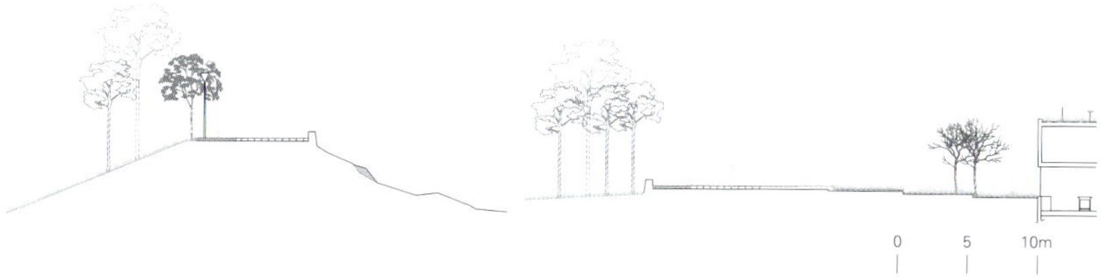
0
5
10m

从多个角度上看，Amagerforbraending都位于一种边缘地带。它位于哥本哈根郊区，也是新娱乐活动的中心。它将当地区域分成两部分，工厂位于一侧，住宅位于另一侧。这个地方声名远播，但鲜有人前往。本案项目的目的在于将上述所有不和谐因素联系起来，在哥本哈根形成一个崭新的统一体。我们想要把它变成一个人们愿意前去旅行的地方。近年来建造的发电厂大多数只具有功能性，外表造价高昂。其外观的主要“用途”是掩盖工厂形象问题及品牌问题严重的事实。我们想做的不只是建造一个外表美观的工厂，更希望能为其增加功能性。追求创造附加功能与追求美观并不冲突。不必非要选择其中一个，两个条件可以同时满足。我们计划筹建一个经济、环保、具有社会盈利性的新废物能源电厂。Amagerforbraending并非孤立的建筑，我们将建筑调动起来，加强建筑与城市之间的联系——在该地区扩大现有活动的范围，将新建筑Amagerforbraending的屋顶变成供哥本哈根市民娱乐的滑雪坡道。

滑雪坡道采用可再生的人造颗粒物质铺设而成，生态环保，打破了室内滑雪场地消耗大量能源的惯例。要到达滑雪坡道需要乘坐工厂烟囱旁边的电梯，在电梯内可以看到工厂内部，一窥工厂内的工作情况，最后到达楼顶高100m处的观景平台，游客可以在这个哥本哈根最高建筑之一的地方尽享美景。烟囱经过改良，每一吨化石燃料中的$CO^2$排放时，烟囱都会释放出直径为30m的烟圈。烟圈是德国艺术设计工作室Realities:United的作品。烟气中的水分在缓慢上升、冷却的过程中凝结，形成烟圈，委婉地提醒人们能源消耗造成的影响，并充当起量尺，直接令哥本哈根公民了解到$CO^2$的排放量，将烟囱这一传统上工业时代的象征转变成未来的象征。夜里，热追光灯将激光投射到烟圈上，形成发光的烟圈艺术。从远处看，整个建筑的表皮被垂直的绿色植物种植块覆盖，像用砖堆砌起来的一样，远看像座山。“山”周围的土地被Topotek 1&Man Made Land公司开发成一个公园，在夏季和冬季作为非正式的体育活动场地。

沿着建筑西侧边缘的山地将模拟建成一个大型滑雪道。公园里的道路系统与整个场地紧密相连，并与周围的住宅区建立了良好的联系。

### The New Waste - to - Energy Plant

Amagerforbraending is in many ways situated on an edge condition. It is a place in the outskirts of Copenhagen, but also the center for new recreational activities. It divides the local area into two, with factories on one side and housing on the other. It is a place you know from afar, but where few people ever go. The aim of the project is to tie all these opposing forces together, forming an identity for a new place in Copenhagen. We want to turn it into a place in itself – a destination where people will travel to. Most of the recently built power plants are merely functional boxes, wrapped in an expensive gift paper. The main "function" of the facade is to hide the fact that factories are having a serious image/branding problem. We want to do more than just create a beautiful skin around the factory. We want to add functionality. The ambition of creating added value in terms of added functionality does not stand in contrast to the ambition to create beauty. It does not have to be either but it can be both. We propose a new breed of waste-to-energy plant, one that is economically, environmentally, and socially profitable. Instead of considering

# 新废物能源电厂

BIG + Realities:United + AKT + Topotek 1&Man Made Land

WASTE IS NOT A WASTE
1 TON WASTE
= 2 MWh DISTRICT HEATING
+ 0.67 MWh ELECTRICITY
ELECTRICITY FOR 550,000 CITIZENS, HEAT FOR 140,000 HOUSEHOLDS
400,000+ TONS OF WASTE/YEAR

THERE IS SOMETHING ROTTEN
IN THE STATE OF DENMARK...
Municipal Waste
Treatment in Denmark
4% LANDFILL
42% RECYCLED
54% WASTE-TO-ENERGY
= WASTE-TO-ENERGY PLANT
Municipal waste treatment
in Western Europe

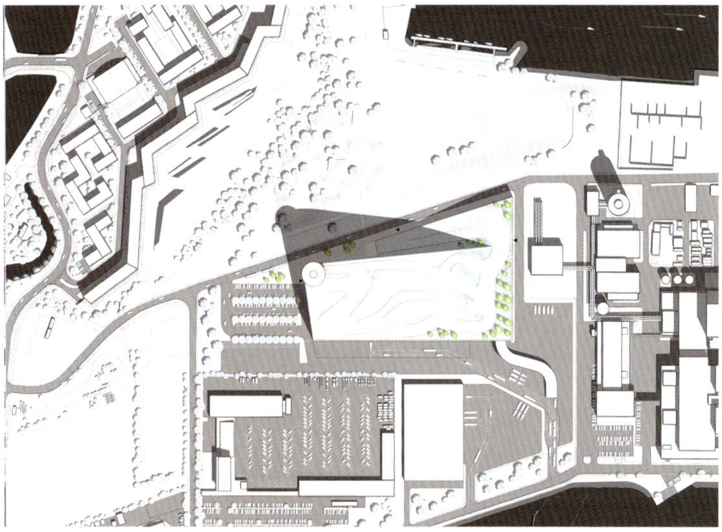

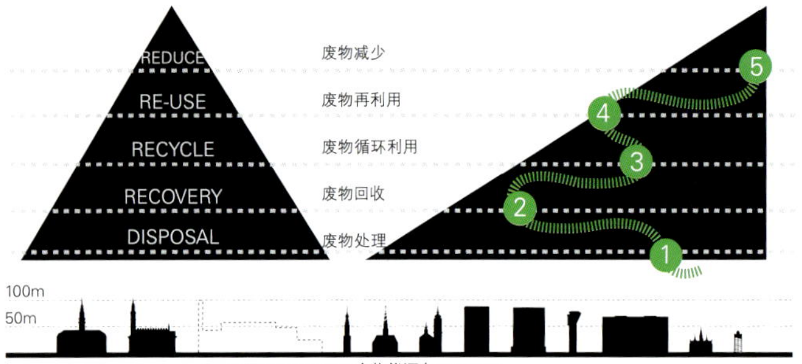

Amagerforbraending废物能源电厂 (2015)
Amagerforbraending WTE plant (2015)

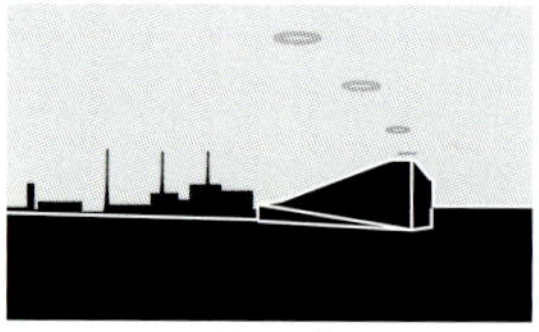
远景 at a distance

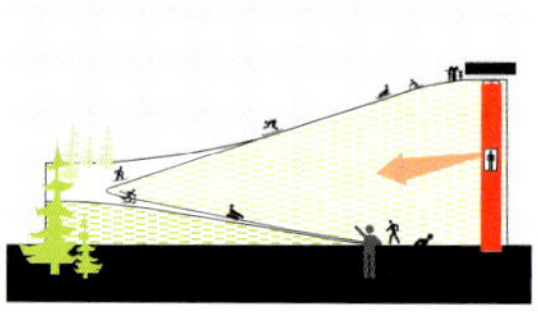
近景 by proximity

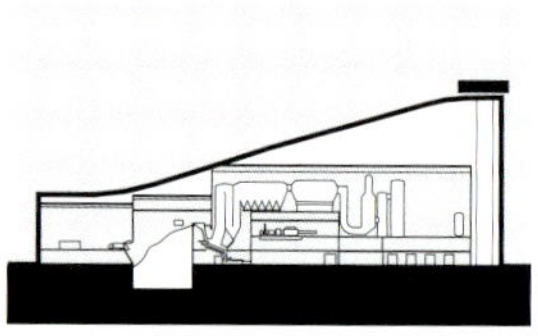
直接接触 direct contact

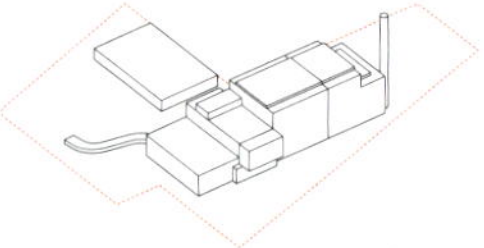
项目 program

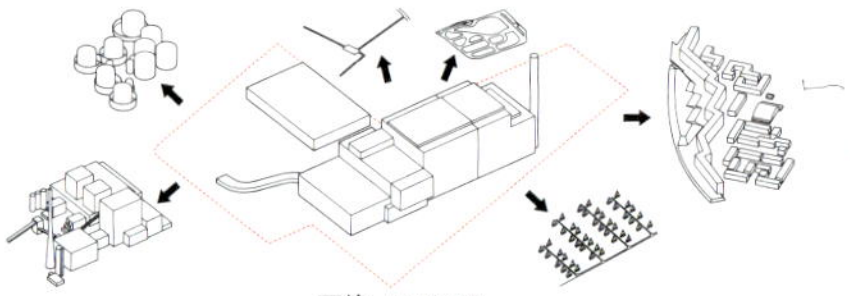
环境 context

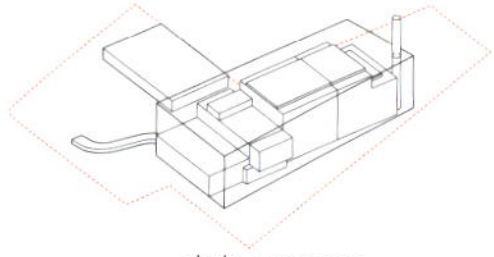
表皮 wrapper

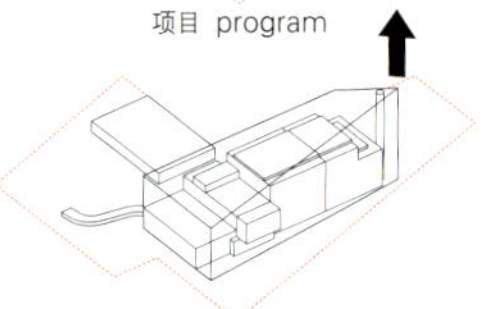
烟囱的整合 integration of smokestack

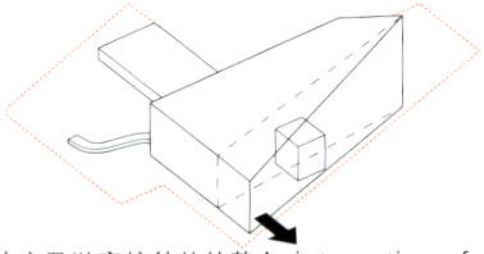
行政中心及游客接待处的整合 integration of administrative & visitor facilities

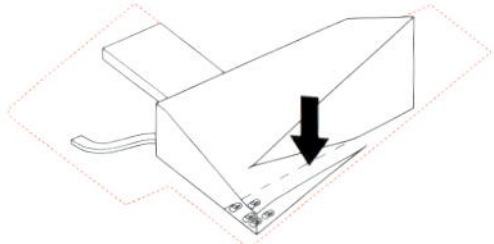
公共连接部分 public connection

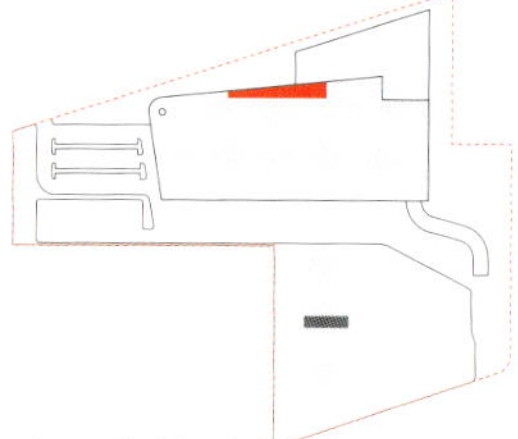
与工厂的联系 relationship to plant

别浪费，要利用 don't waste it, use it

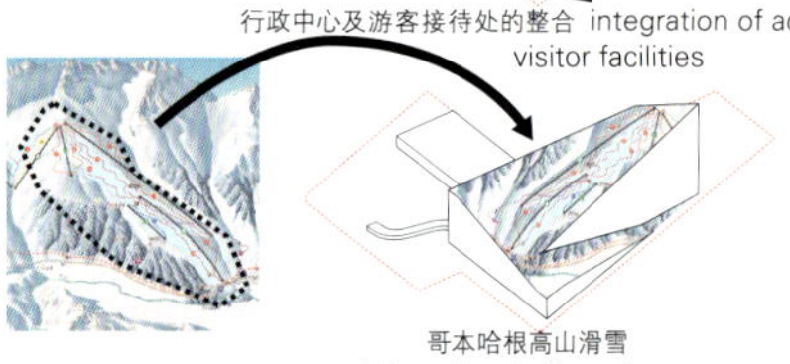
哥本哈根高山滑雪
alpine sking in Copenhagen

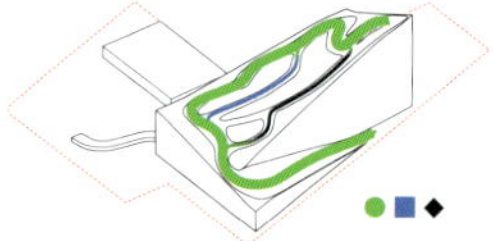
三个滑雪道 three pistes

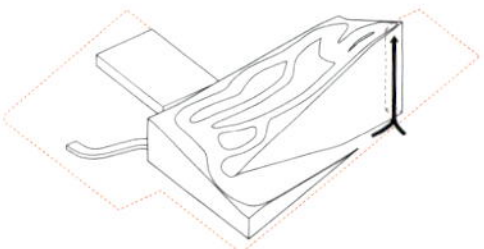
滑雪坡道入口 slope access

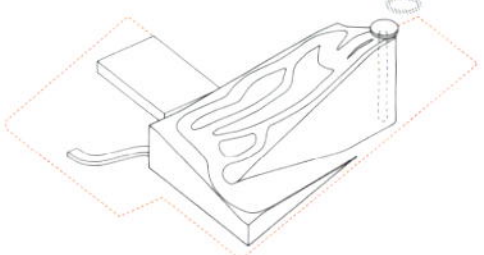
烟圈 smoke rings

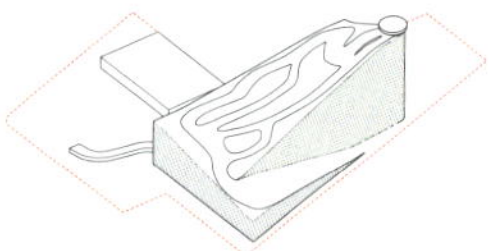
绿色墙体 green walls

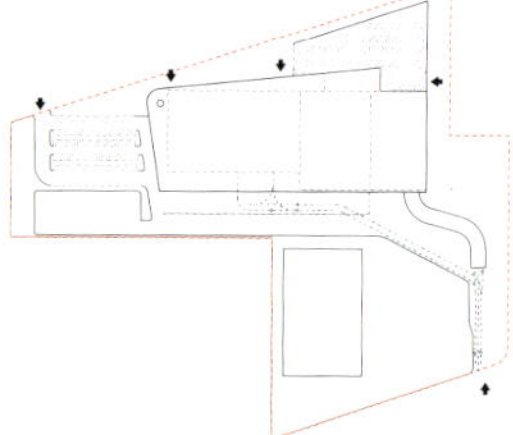
停车场及主要通道 parking and main access

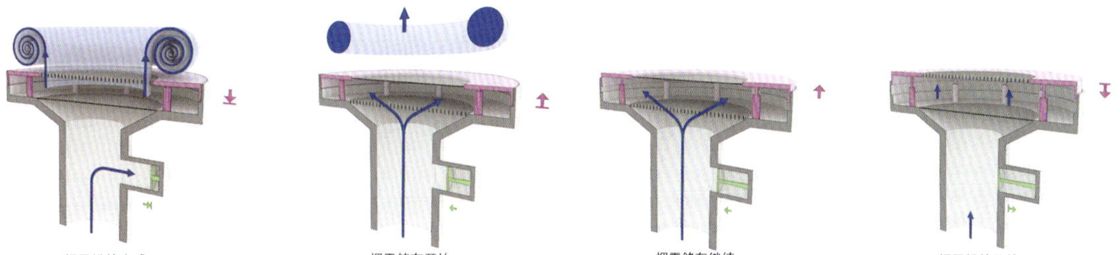


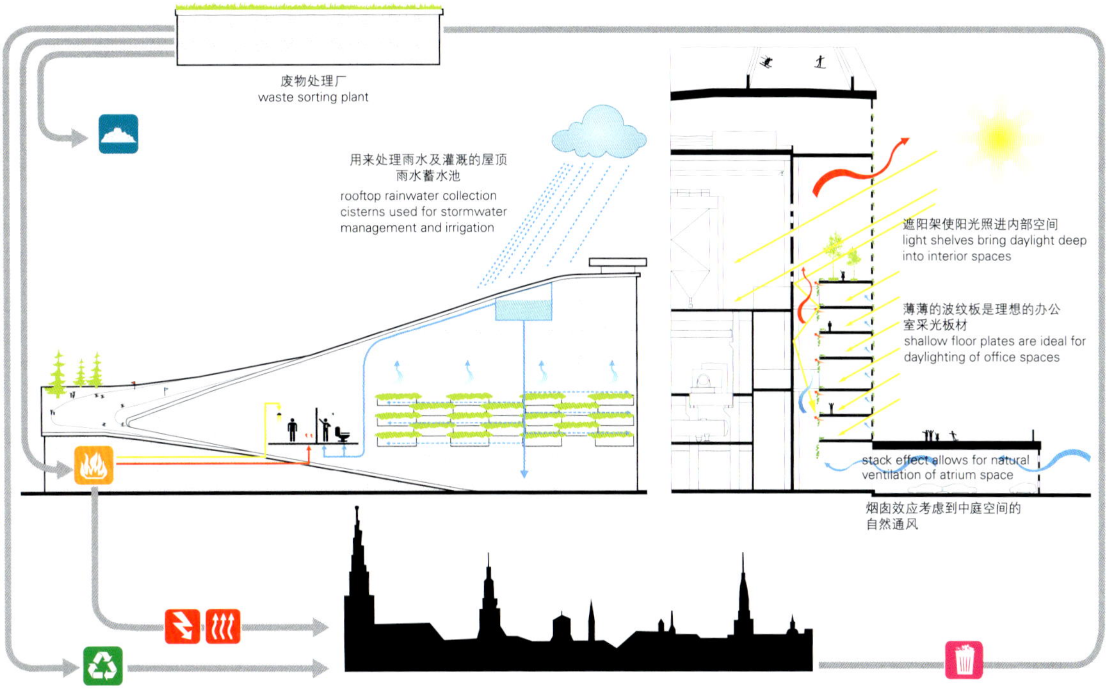

项目名称：Waste-to-Energy Plant
建筑师：BIG-Bjarke Ingels Group
项目主管：Brian Yang
项目团队：Jelena Vucic, Alina Tamosiunaite, Armor Gutierrez, Maciej Zawadzki, Jakob Lange, Andreas Klok Pedersen, Daniel Selensky, Gül Ertekin, Xing Xiong, Sunming Lee, Long Zuo
烟圈发电机：Realities:United
立面及结构顾问：AKT
景观：Topotek 1 / Man Made Land
甲方：Amagerforbraending
地点：Amager, Copenhagen, Denmark
建筑面积：95,000m²
景观面积：90,000m²
立面面积：74,000m²
屋顶+滑雪坡道面积：32,000m²
行政办公+游客中心建筑面积：6,500m²
获奖情况：1st prize winning project

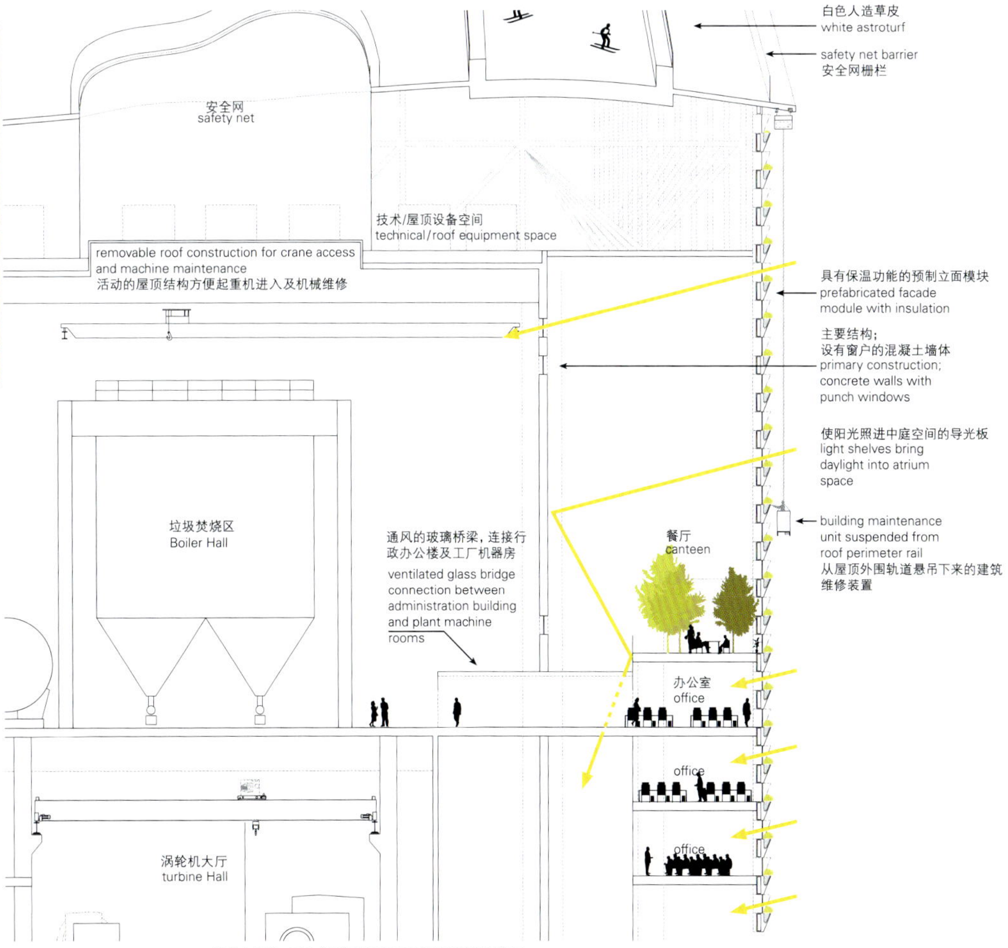

行政办公楼、工厂主要结构及屋顶之间关系的剖面详图
detail section of relationship between administrative building, main plant volume, and roof

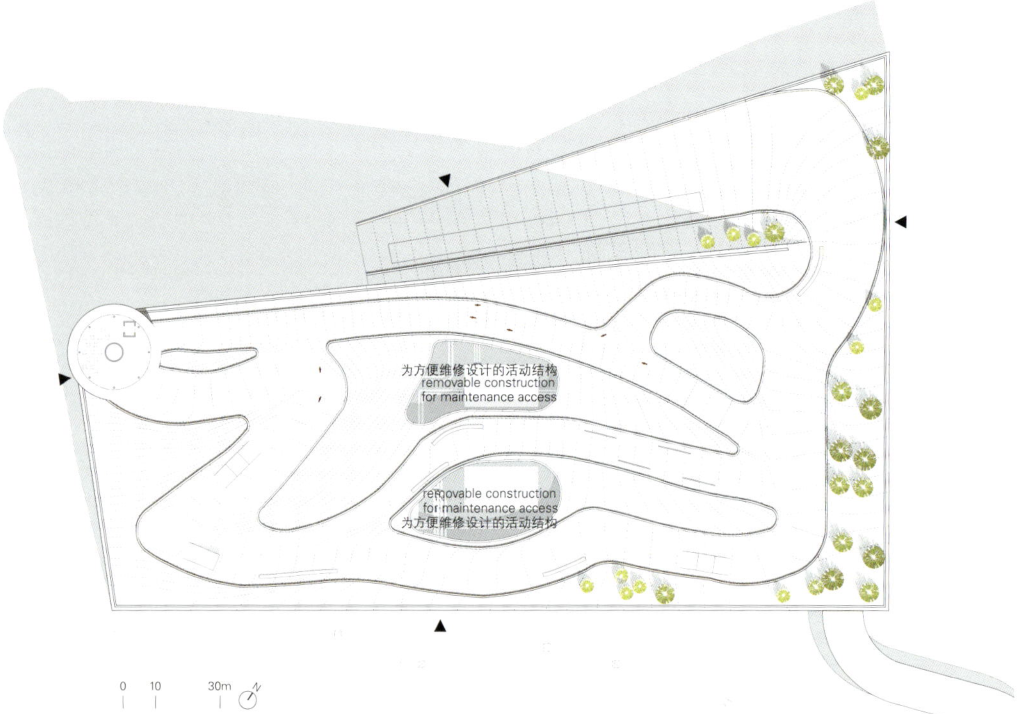


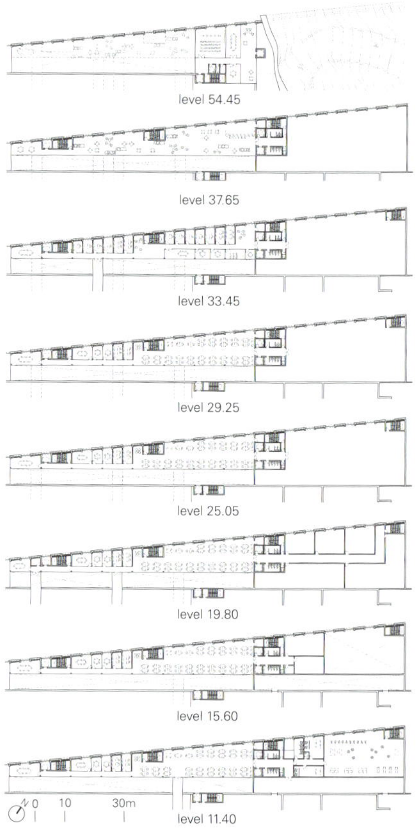


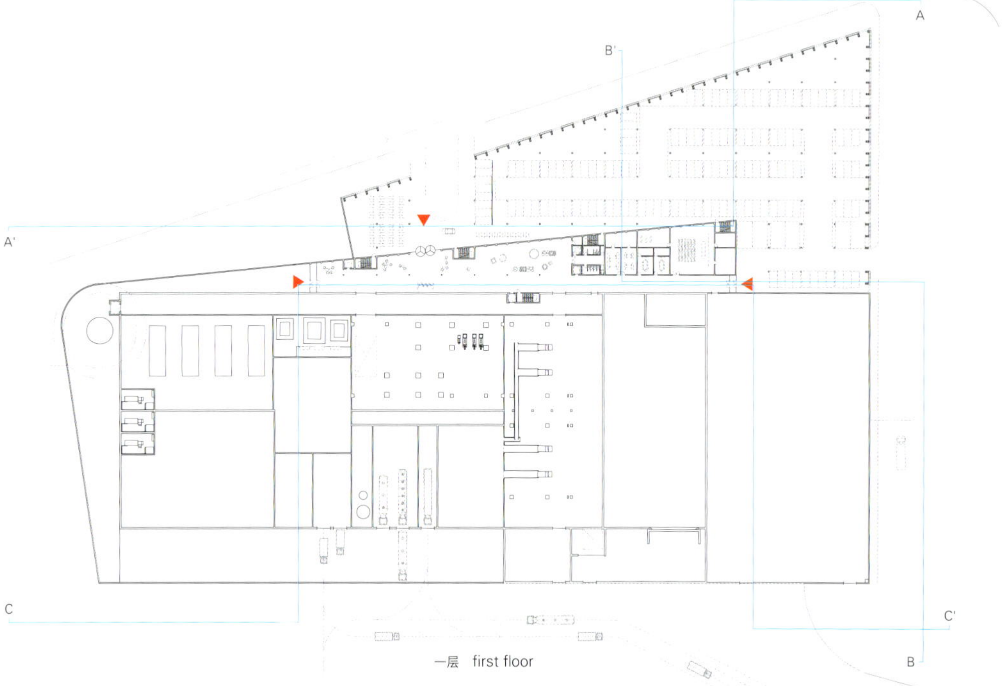


一层 first floor

Amagerforbraending as an isolated object, we mobilize the architecture and intensify the relationship between the building and the city – expanding the existing activities in the area by turning the roof of the new Amagerforbraending into a ski slope for the citizens of Copenhagen.

The slope is ecological using a recycled synthetic granular, upending the convention of the energy intensive indoor ski resort. Access to the slopes is via an elevator along the plant's smokestack providing views into the plant, giving glimpses of its inter-nal workings finally reaching an observation platform 100m above giving sightseers an unobstructed view from one of the tallest structures in Copenhagen. While the smokestack is modified to puff smoke rings of 30m in diameter whenever 1 ton of fossil $CO_2$ is released. These smoke rings which are the brainchild of Germany-based art studio Realities:United will form due to the condensation of water in the flue gases as they slowly rise and cool, serving as a gentle reminder of the impact of consumption and a measuring stick that will allow the common Copenhagener to grasp the $CO_2$ emission in a straightforward way turning the smokestack traditionally the symbol of the industrial era into a symbol for the future. At night, heat tracking lights are used to position lasers on the smoke rings into glowing artworks.

From a distance, the entire building is wrapped in a vertical green facade formed by planter modules stacked like bricks turning it into a mountain from afar. The surrounding topography is sculpted by Topotek 1&Man Made Land to form a park which offers informal sport activities in summer and winter.

Along its western border a field of hills will mimic a mogul piste. A path system throughout the park interconnects the whole site and creates good connections to the neighboring residential area.

BIG-Bjarke Ingels Group

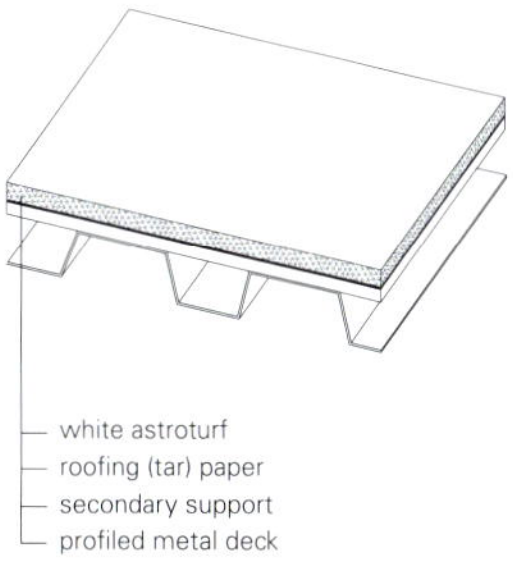


典型的空置屋顶构造
typical non-occupied roof build-up

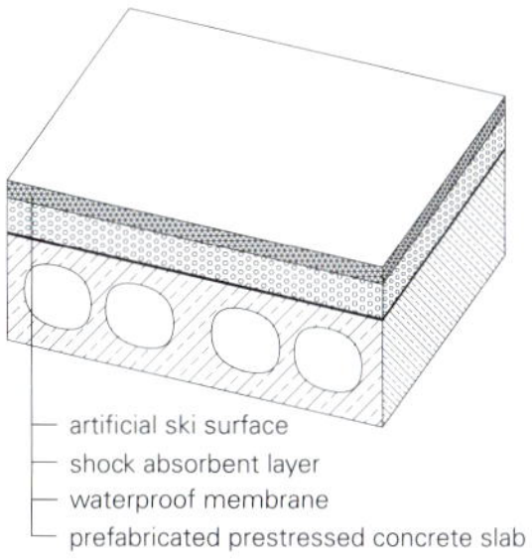


滑雪场表面屋顶构造
ski surface roof build-up

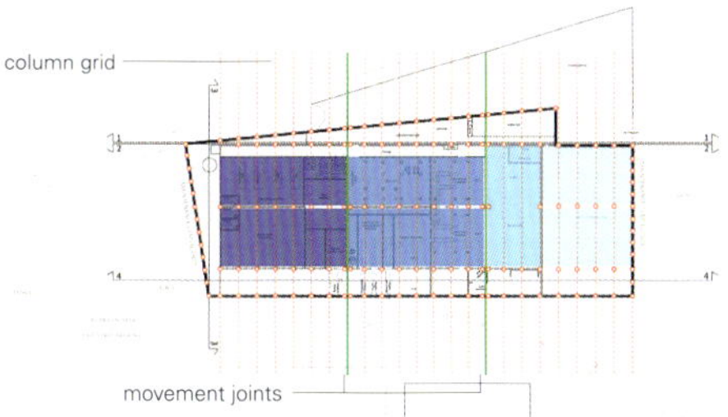


柱网及移动接缝的结构示意图
structural layout showing column grid and movement joints

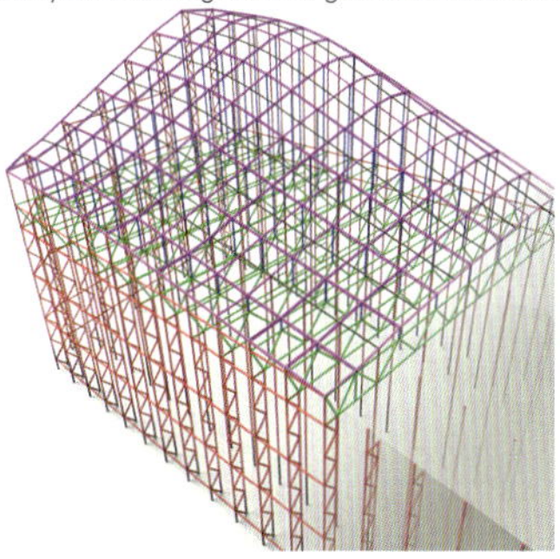

三维透视结构效果图
3D perspective structural rendering

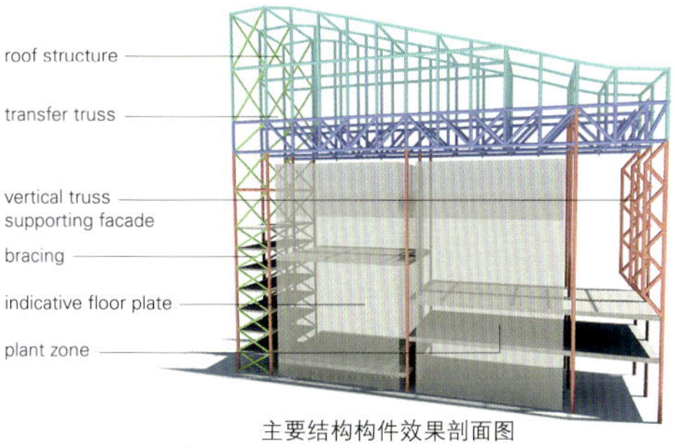


主要结构构件效果剖面图
rendered section showing prinicpal structural elements

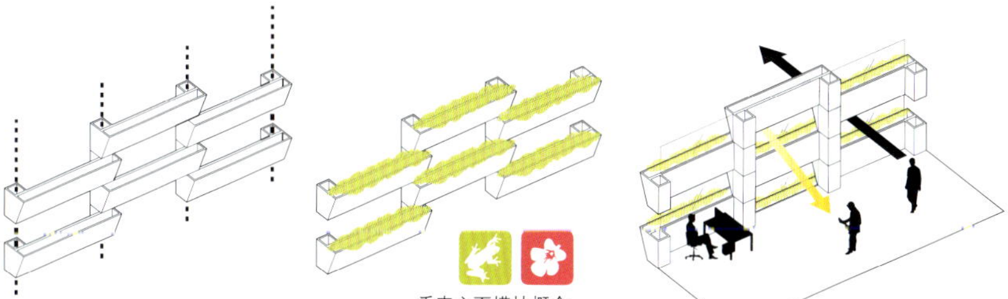

垂直立面模块概念
vertical facade module concept

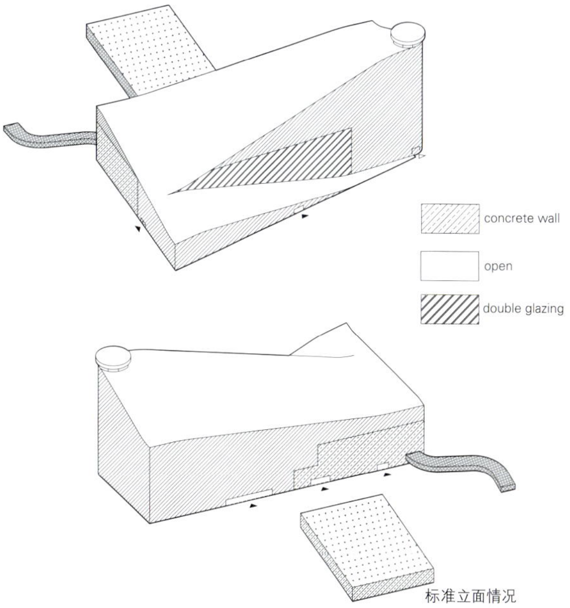


标准立面情况
typical facade conditions

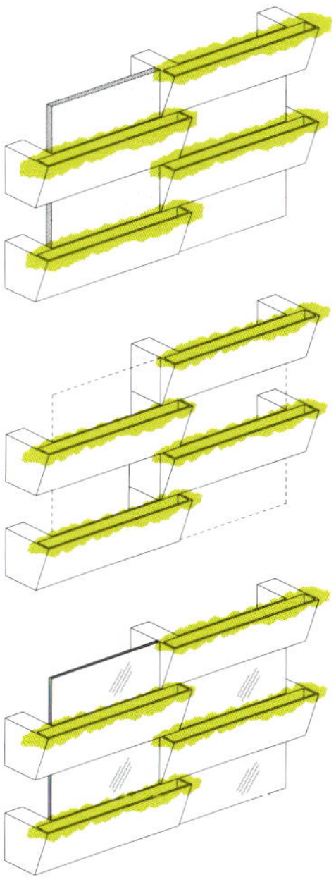

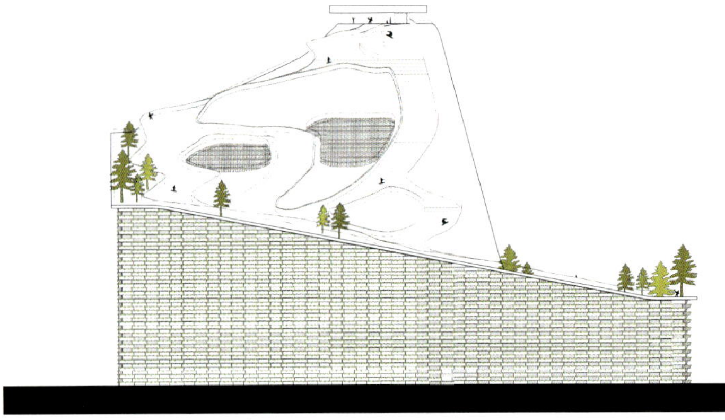

东北立面　north-east elevation

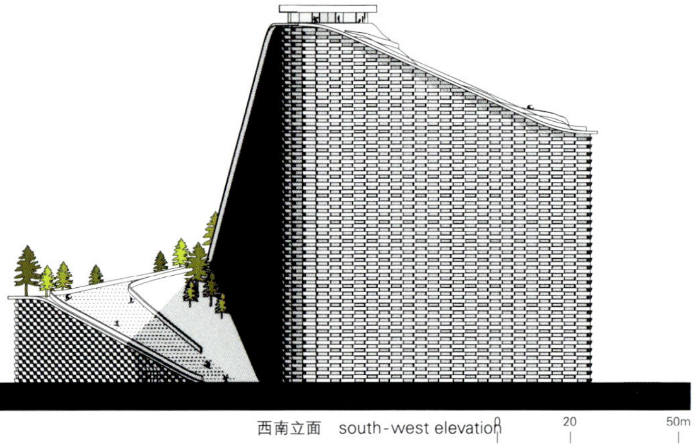


西南立面　south-west elevation

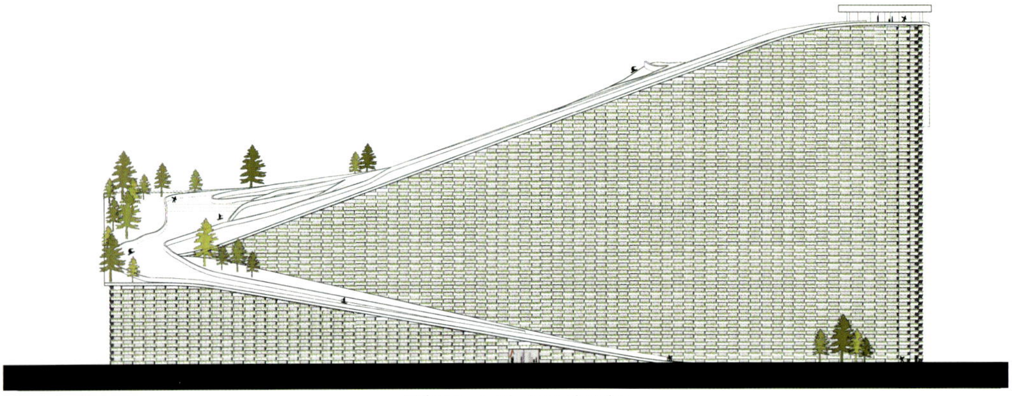

西北立面　north-west elevation

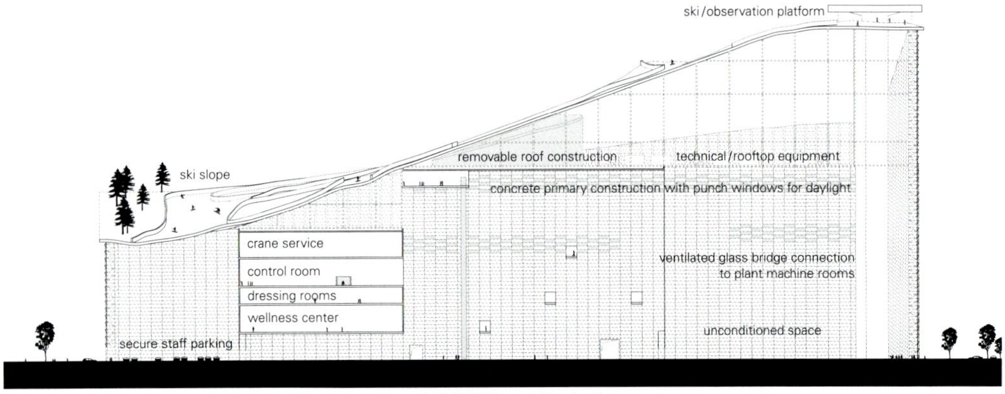


A-A'剖面图　section A-A'

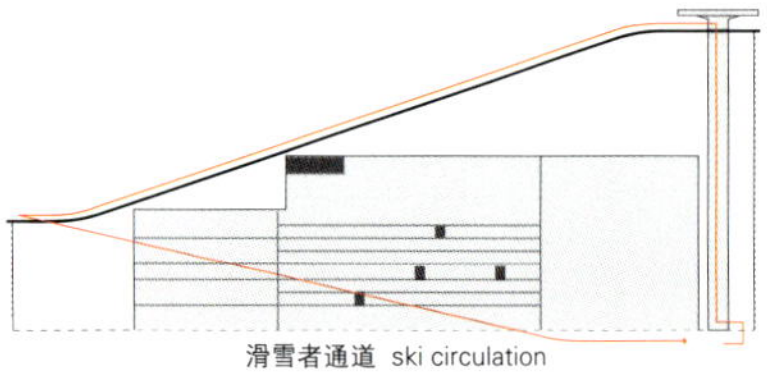

滑雪者通道　ski circulation

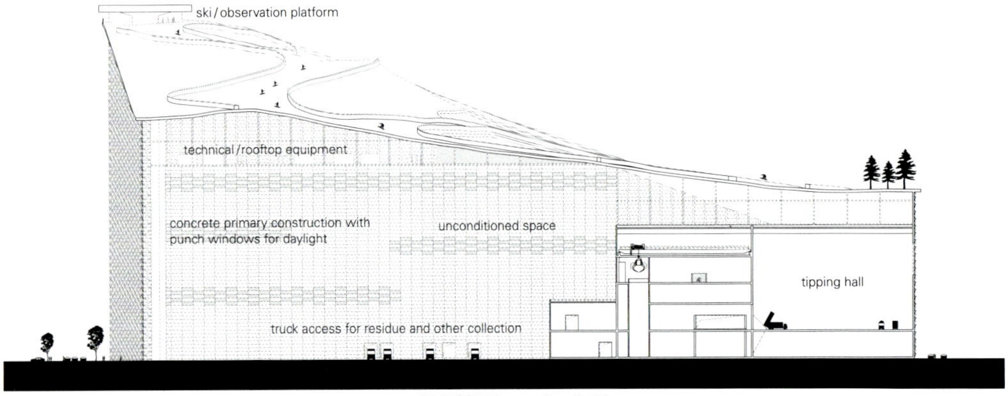


B-B'剖面图　section B-B'

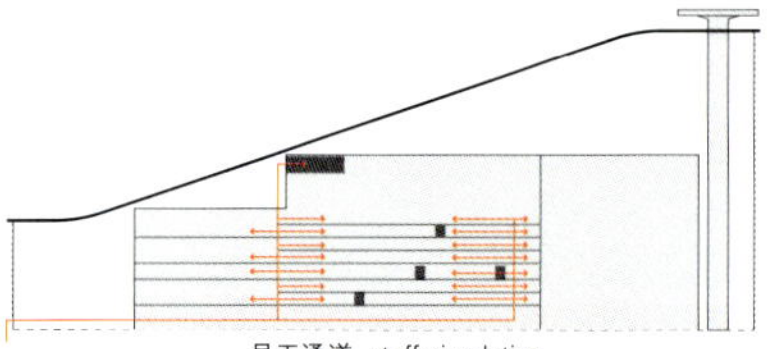

员工通道　staff circulation

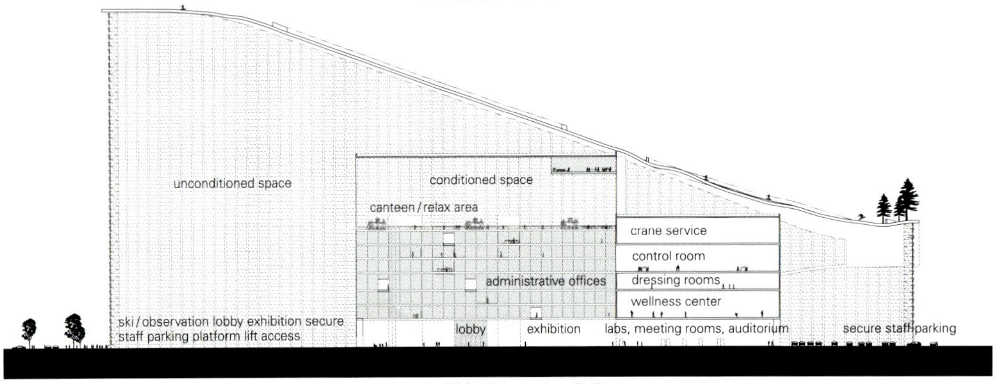


C-C'剖面图　section C-C'

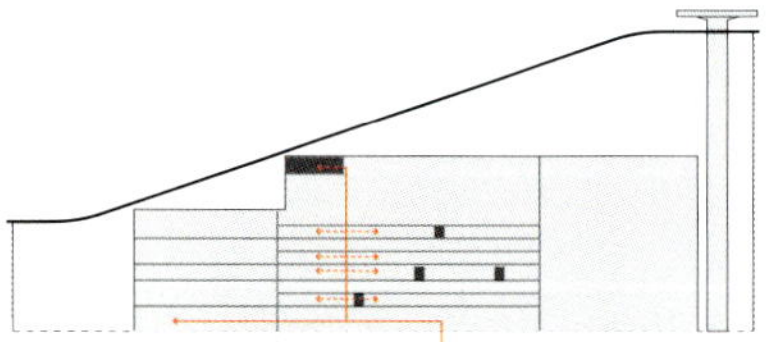

游客通道　visitor circulation

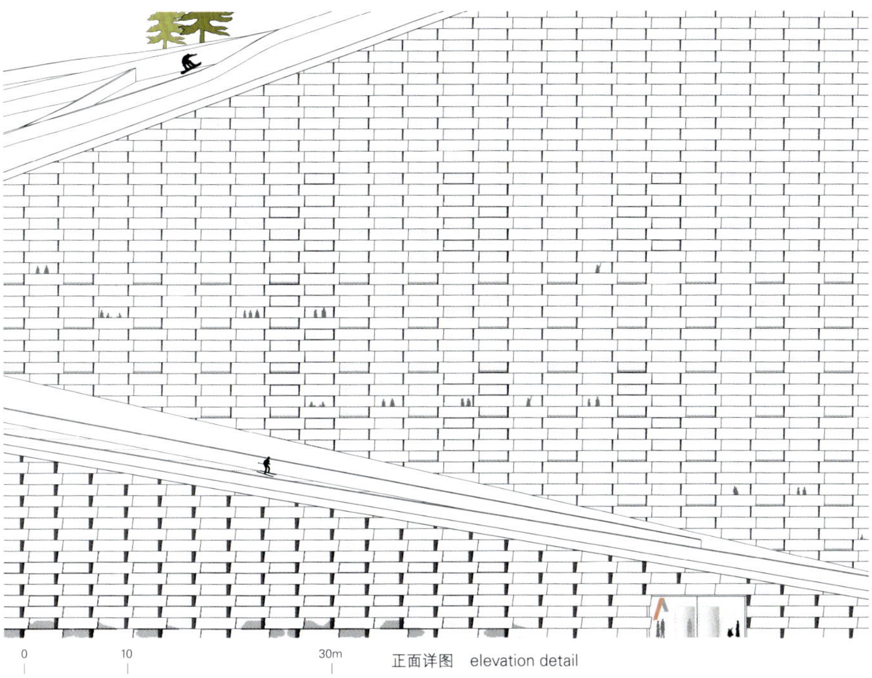


正面详图　elevation detail

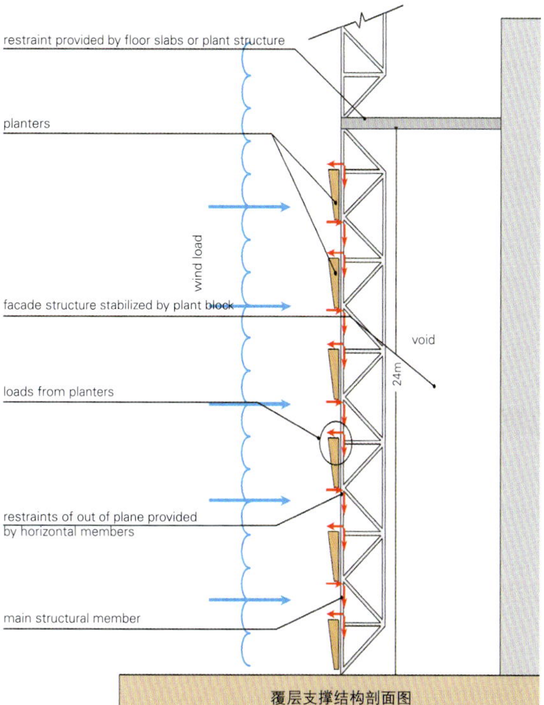


覆层支撑结构剖面图
cladding support structure section

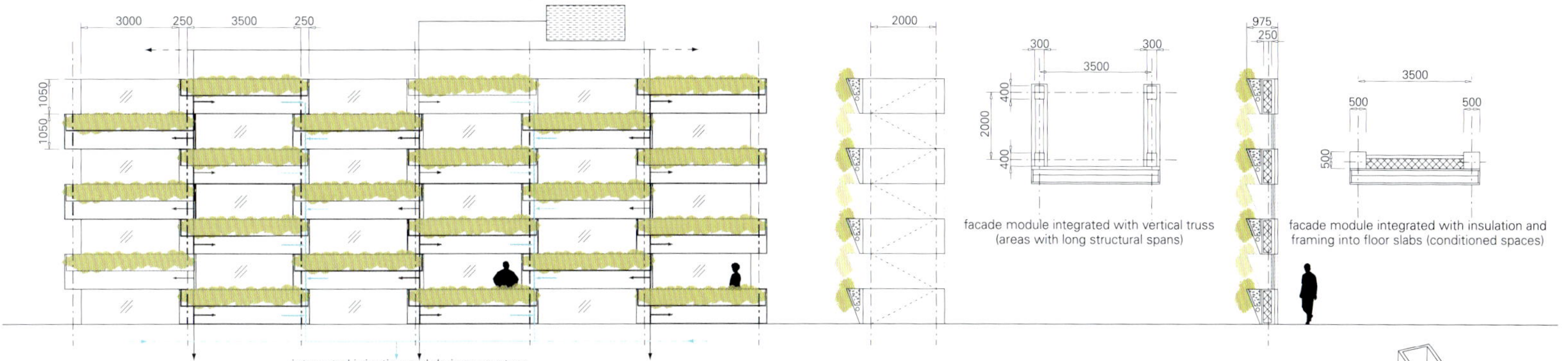


标准立面模块尺寸　typical facade module dimensions

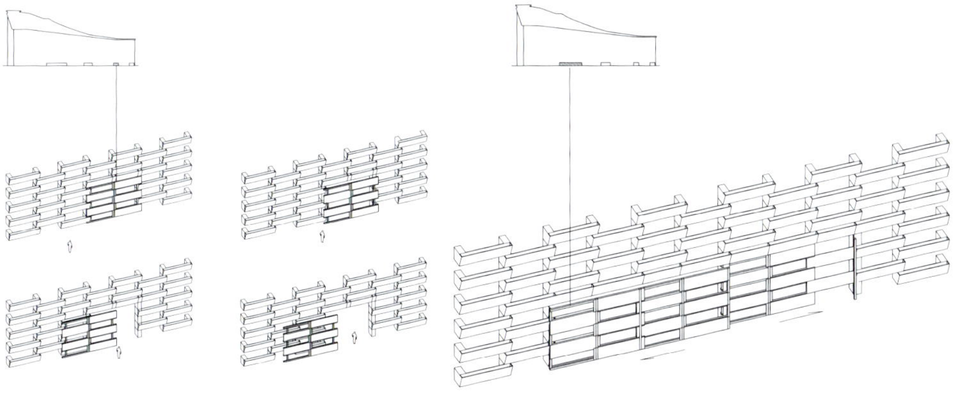

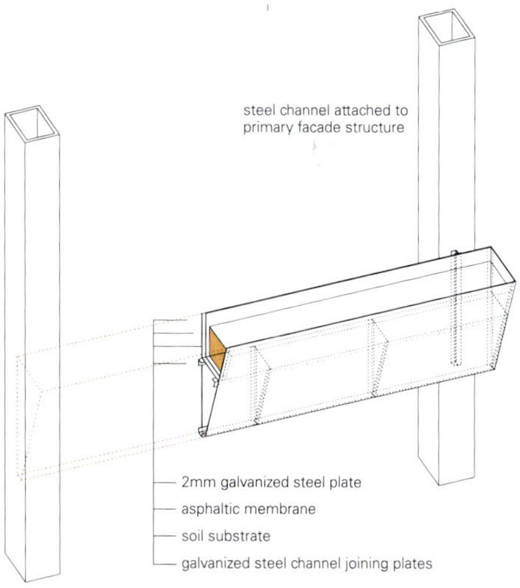

格陵兰岛国家美术馆坐落于一个陡峭的斜坡上，可俯瞰格陵兰风景最优美的峡湾，面积3000m²，它将是格陵兰人民文化及建筑的图标。这座新博物馆将融格陵兰的历史及当代艺术为一体，成为一座动感的建筑。

努克的丹麦功能主义建筑基本以方盒子为主，忽略了格陵兰岛独特的自然环境。因此，我们要将国家美术馆设计成一个无论在实体上还是视觉上都与大自然和谐统一的建筑，就像格陵兰岛上的生活一样，成为自然的共生体。我们创造了一个简单且具有功能性和象征性的形体，将完美的圆形和当地的地形相结合，创造出一个独一无二的混合形态，既包括抽象特性，又具有地域特性。

新美术馆可以看成是一个完美的几何圆形投影在陡峭的山坡上，像一个四合院建筑，将纯粹的几何图形与周边景观的敏感适应性相结合。景观的三维特征围绕博物馆的焦点——雕塑公园，创造出一个保护环。在雕塑公园中，参观者、全体员工及展览品与文化、自然内外融合在一起。

斜坡使雕塑公园向城市及风景开放，塑造出雕塑公园和博物馆的功能。建筑粗糙的白色混凝土立面会随时间的推移而生铜绿并适应当地的气候；而建筑内部圆形的玻璃面简单而精致，与粗犷的自然形成鲜明对比，称颂这优美的景色。

美术馆的环形结构使其可以被灵活划分成形状大小各异的展厅，为博物馆艺术创造出一种独特的框架。游客在建筑正面乘电梯穿过有屋顶的通道进入大厅，可以180° 观赏到雕塑公园及峡湾全景，同时可以体验博物馆的公共功能，如售票台、更衣室、精品店、咖啡厅。

新美术馆将调动整个区域，在峡湾水边创建更多的活动，通过这种随地势起伏而建造的博物馆与周边地区联系紧密。当地居民及游客可以把这座美术馆当作一座雕塑或一件大地艺术来欣赏其清晰的形状。

**Greenlands National Gallery**

Located on a steep slope overlooking the most beautiful of Greenland's fjords, the 3000m² National Gallery will serve as a cultural and architectural icon for the people of Greenland. The new museum will combine historical and contemporary art of the country in one dynamic institution.

The Danish functionalistic architecture in Nuuk is typically square boxes which ignore the unique nature of Greenland. We therefore propose a national gallery which is both physically and visually in harmony with the dramatic nature, just like life in Greenland is a symbiosis of the nature. We have created a simple, functional and symbolic shape, where the perfect circle is supplied by the local

# 格陵兰岛国家美术馆

BIG + TNT Nuuk + Rambøll Nuuk + Arkitekti

项目 program

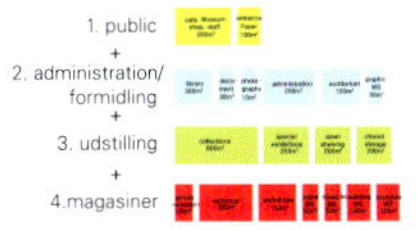

四种规划 four kinds of program

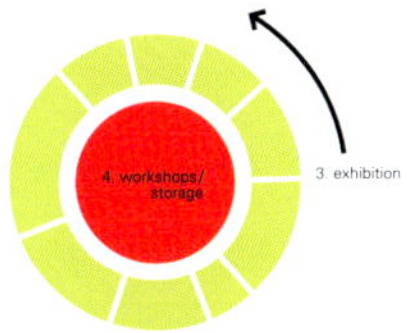

环绕工作室/仓库而建的展厅
exhibition around workshops / storage

行政中心及扩展区
administration and outreach

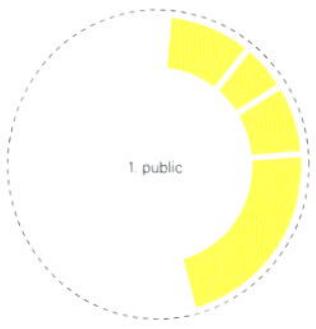

公共场所 public

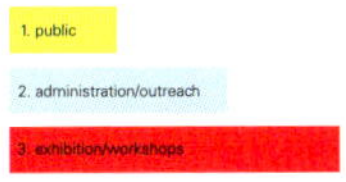

三个层面 three layers

融合 melting

三个环 triple loop

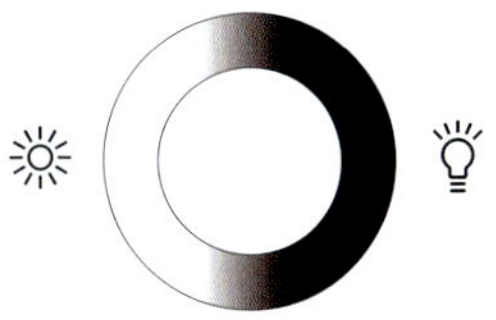

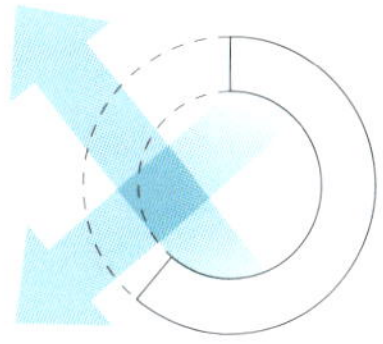

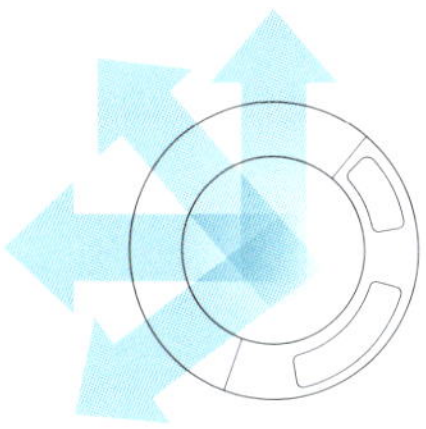

日光 daylight

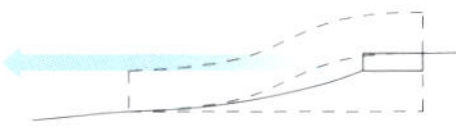
从行政中心观赏到的景色
outlook from administration

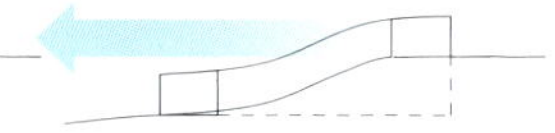
从门厅观赏到的景色
outlook from foyer

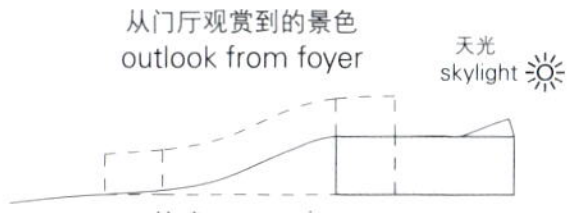

扩建 extension

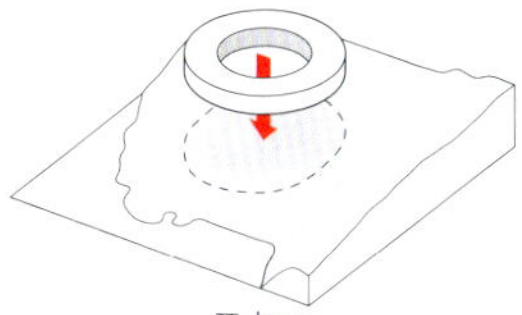
环 loop

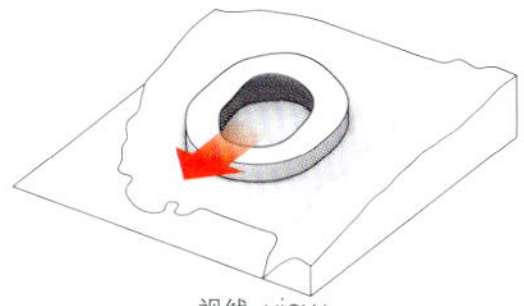
视线 view

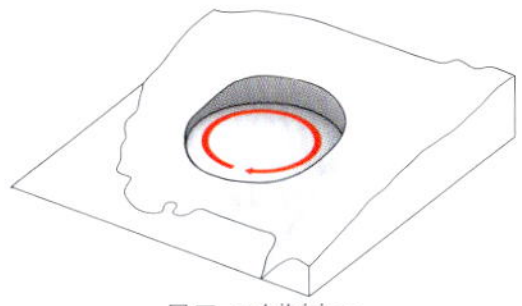
展厅 exhibition

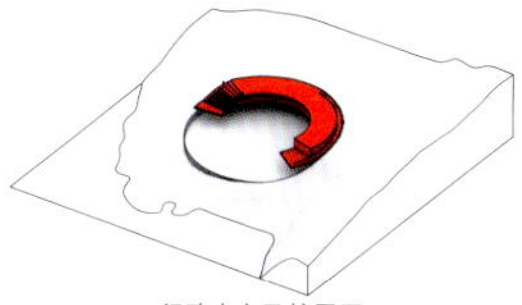
行政中心及扩展区
administration and outreach

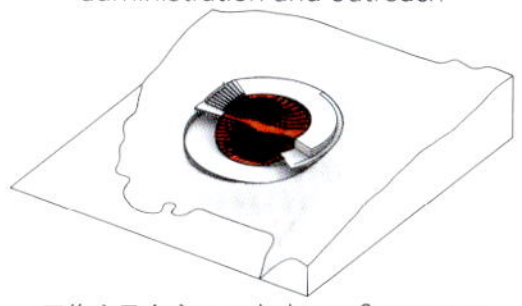
工作室及仓库 workshops & storage

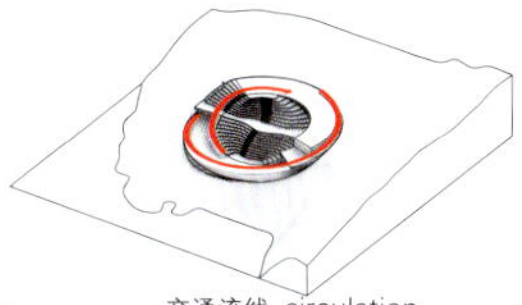
交通流线 circulation

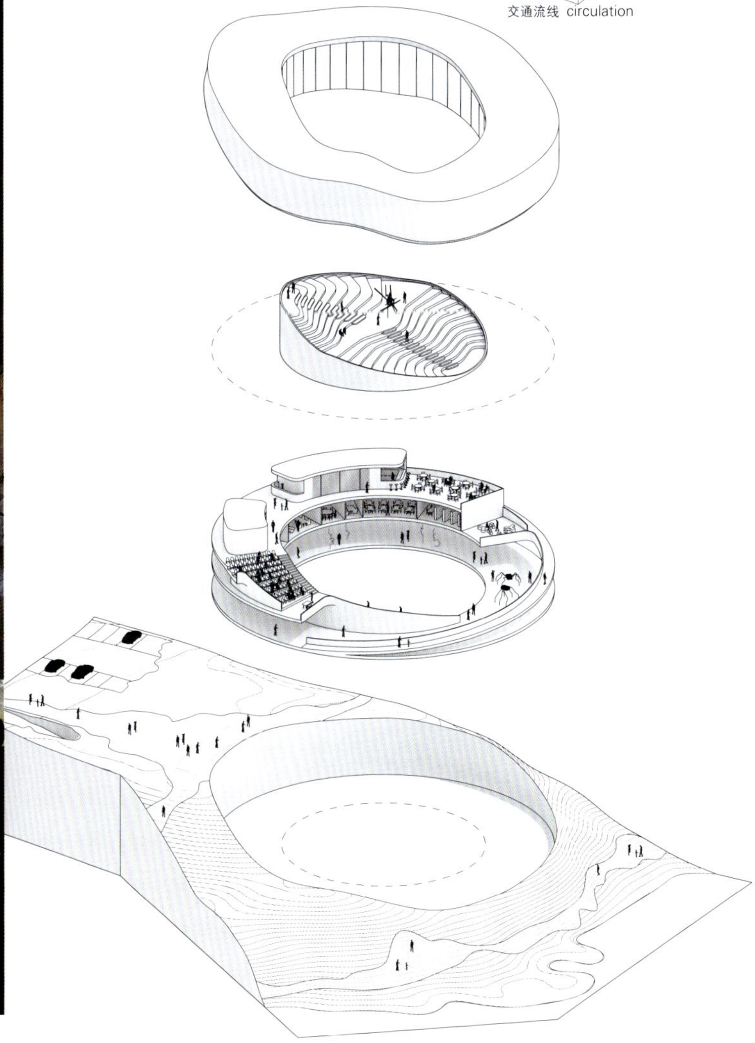

0
20
50m
N

topography which creates a unique hybrid between the abstract shape and the specific location.

As a projection of a geometrically perfect circle on to the steep slope, the new gallery is conceived as a courtyard building that combines a pure geometrical layout with a sensitive adaption to the landscape. The three-dimensional imprint of the landscape creates a protective ring around the museum's focal point, the sculpture garden where visitors, personnel, exhibition merge with culture and nature, inside and outside.

The slope opens up the sculpture garden towards the city and the view, framing both the sculpture garden and museum functions. A rough looking external facade of white concrete will patinate over time and adjust to the local weather, while the circular inner glass facade will consist of a simple and refined frame which contrasts the rough nature and compliments the beautiful view.

The circular shape of the gallery enables a flexible division of the exhibition into different shapes and sizes, creating a unique framework for the museum's art. Visitors access to the exhibition happens through a covered opening created by a slight lift in the facade into a lobby with a 180 degree panorama view towards the sculpture garden and the fjord as well as access to the common museum functions, including ticket counters, wardrobe, boutique and a cafe.

The new gallery will create more activities at the waterfront by attracting the whole area is interconnected by a path which like the museum, forms after the shifting inclinations of the terrain. The locals and visitors will be able to admire the clear shape of the gallery which appears as a sculpture or a piece of land-art.

BIG-Bjarke Ingels Group

项目名称：Greenland National Gallery
建筑师：BIG-Bjarke Ingels Group
合伙负责人：Bjarke Ingels, Andreas Pedersen
项目主管：Jakob Henke
项目团队：Daniel Selensky, Ji-young Yoon, Gul Ertekin, Aleksander Tokarz, Alessio Zenaro, Johan Cool, Nicklas Antoni Rasch
合作者：TNT Nuuk, Rambøll Nuuk, Arkitekti, MIR, Glessner Group
甲方：Nunatta Eqqumiitsulianik Saqqummersitsivia
地点：Nuuk, Greenland
建筑面积：3,000m²
获奖情况：1st prize winning project

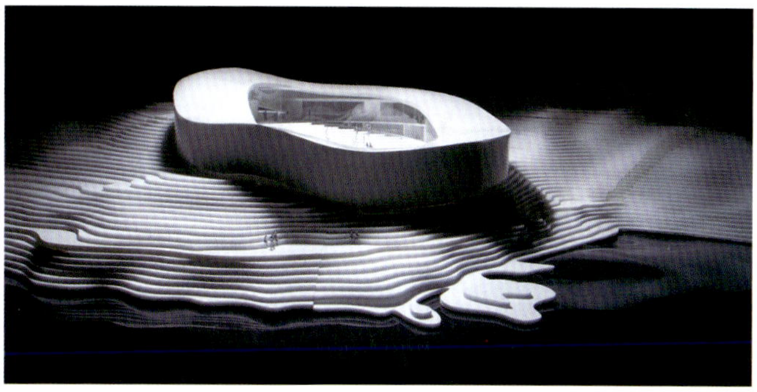

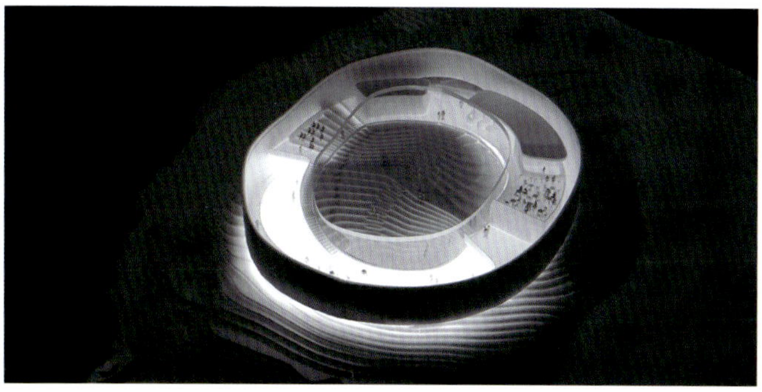

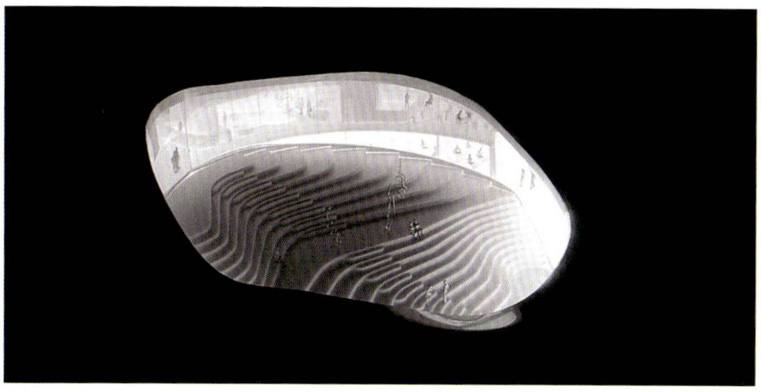

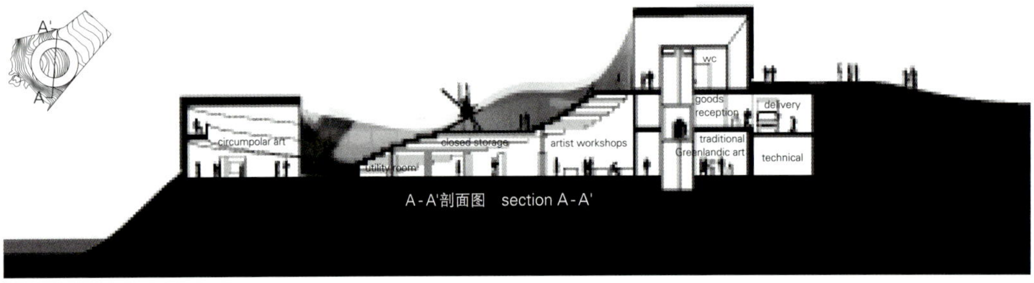

A-A'剖面图　section A-A'

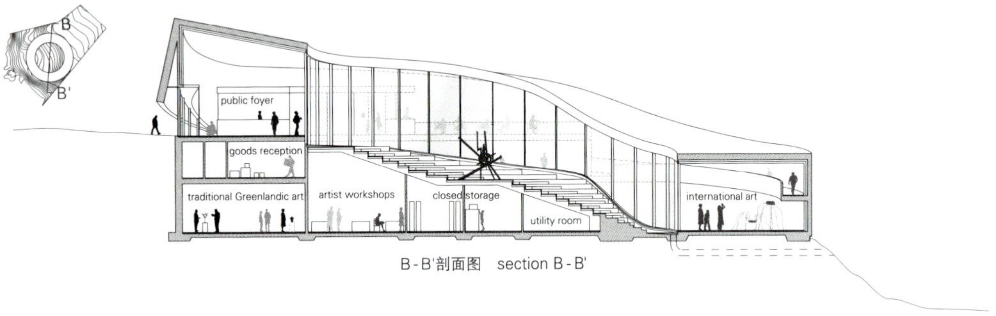

B-B'剖面图　section B-B'

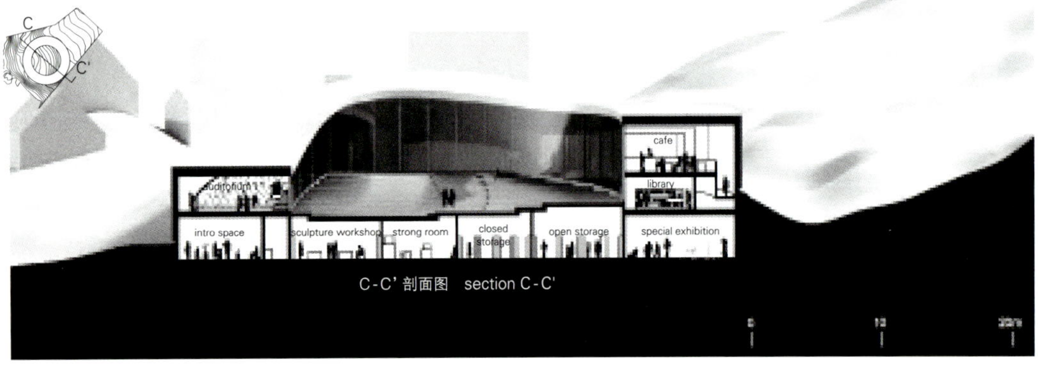

C-C' 剖面图　section C-C'

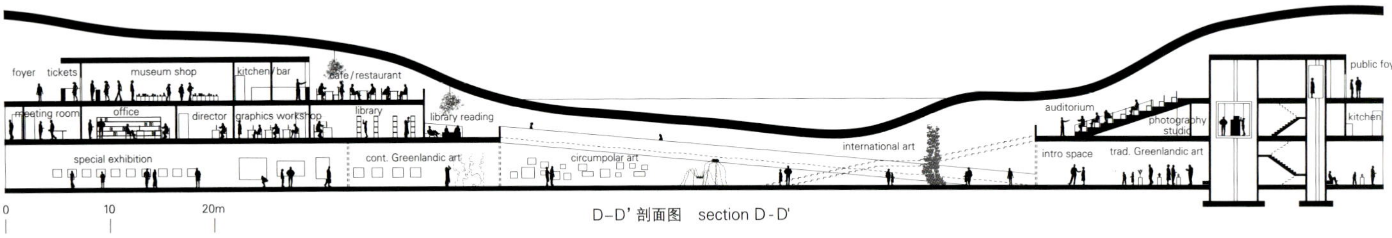

D-D' 剖面图　section D-D'

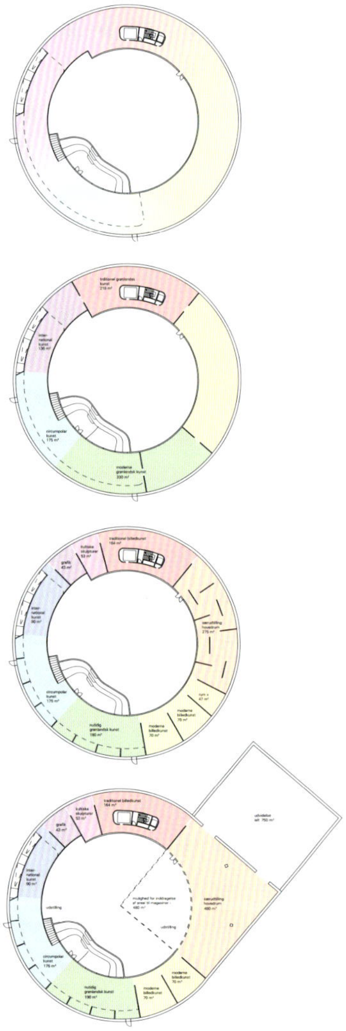

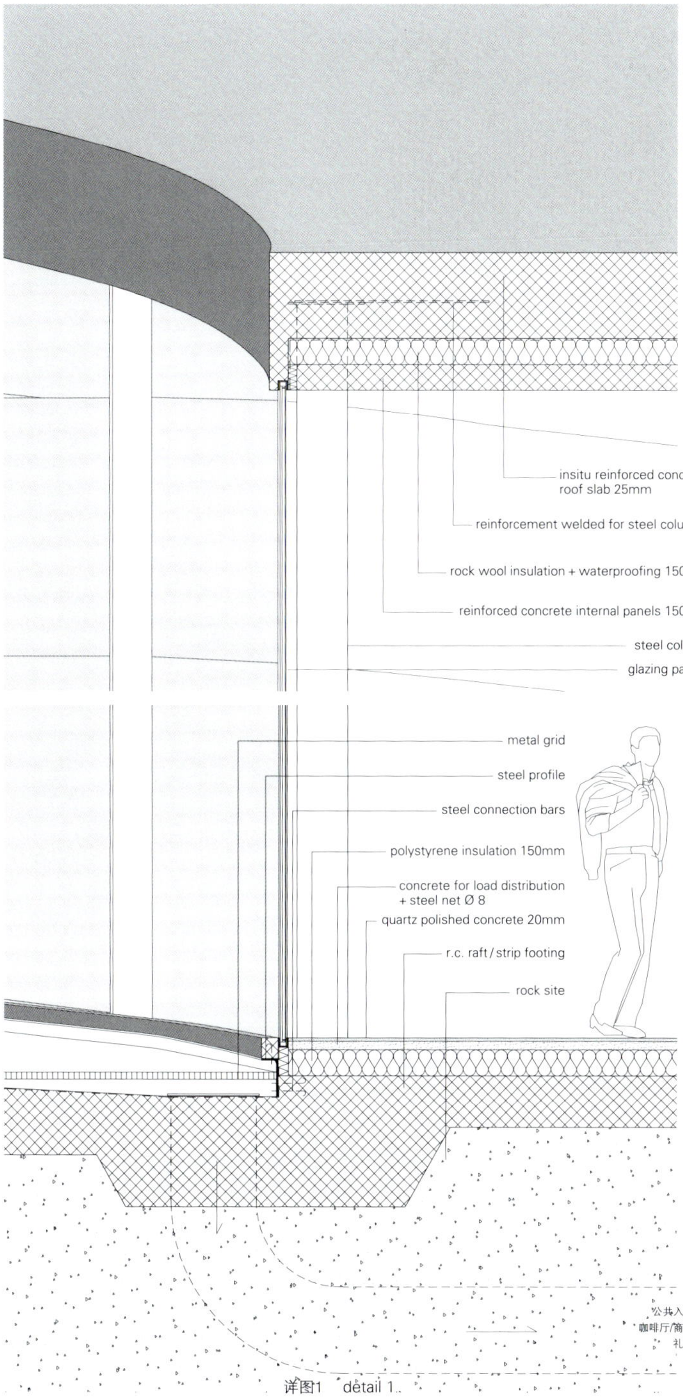


详图1 detail 1

公共入口 public entrance 178m²
咖啡厅/商店 cafe/shop 208m²
礼堂 auditorium 198m²

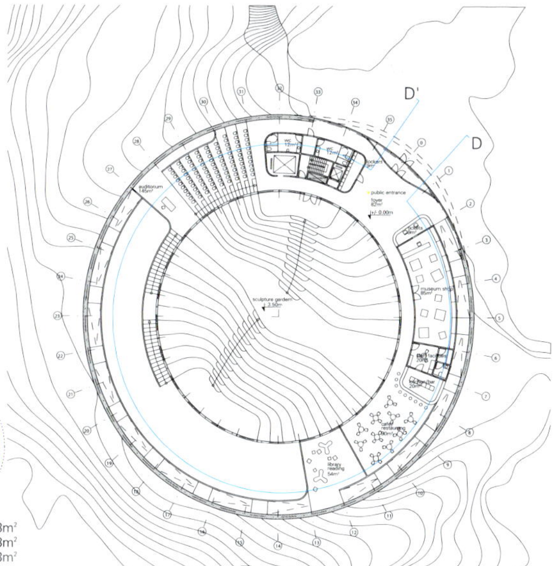


一层 first floor

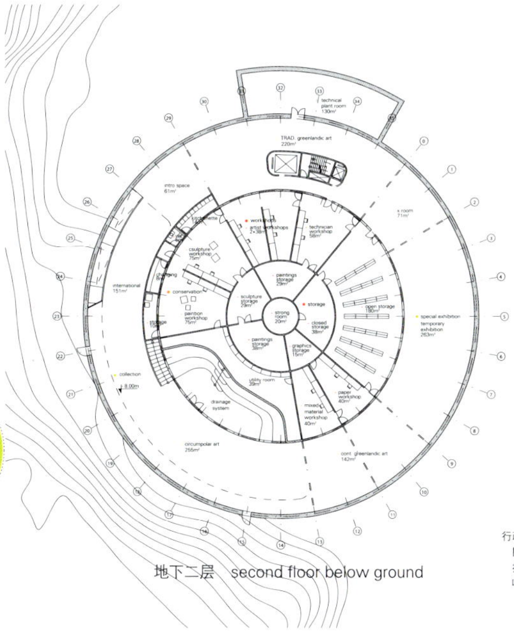

收藏区 collection 829m²
展区 exhibitions 334m²
工作室 workshop 171m²
管理区 conservation 230m²
仓库 storage 349m²
技术室 tech room 169m²

地下二层 second floor below ground

行政中心 administration 178m²
图书室 library 208m²
扩展区 outreach 198m²
收货区 goods reception 109m²

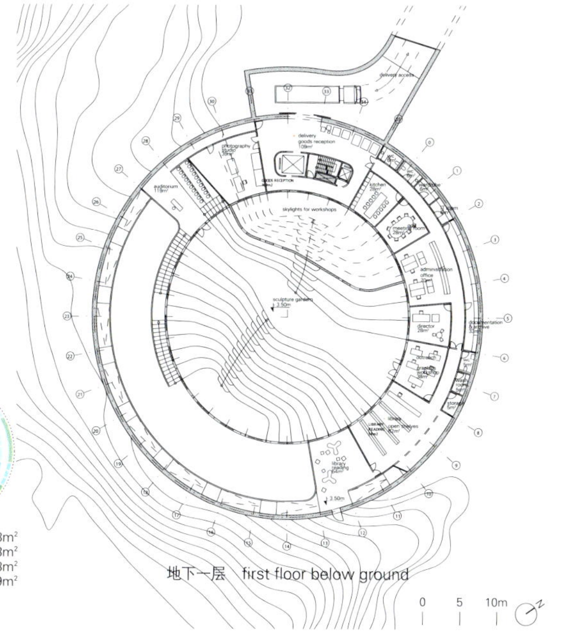


地下一层 first floor below ground

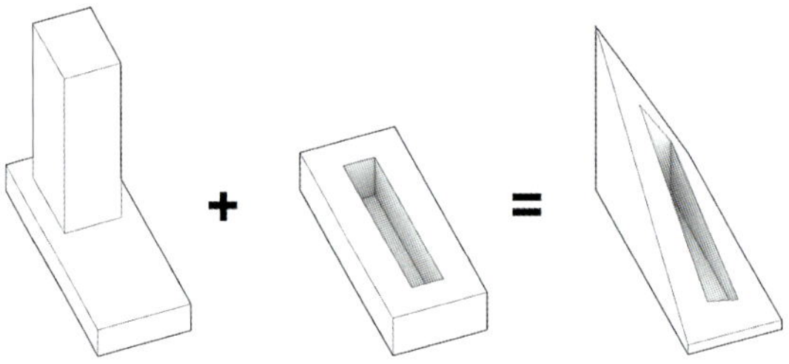


# 西57街

这座建筑是由BIG建筑师事务所在北美的第一个项目。它的规划是基于一个文化和商业的规划，由600多个不同规模的住宅单元组成。该建筑正在为争取LEED金奖认证而努力。

西57街住宅向纽约引进了一个全新的住宅模式，这将给曼哈顿的天际线增添一个吸引人的扭曲线条。

BIG建筑师事务所回顾过去以审视曼哈顿的系统规则，并且正在引进欧洲的建筑模式：周边街区。带着一种高效的布局和亲密的感觉，周边街区能够满足密集性和安全性的要求。

在这个项目里，我们已经把它融入到传统的曼哈顿高层建筑中，开创了一种集两种优点于一身的独一无二的形态：庭院建筑既通风又紧密，同时还有摩天大楼那种令人惊叹的视野。这座建筑的形式会根据观赏者的观看角度而改变，从西部的高速公路上看，它像是一个金字塔，从西58街上看，像是一个神奇的玻璃塔顶。

降低街区的三个拐角，抬高建筑的东北部，从庭院里可以看到哈得孙河的开阔景色，西下的阳光还能照射到街区里。这个庭院对于住户来说还是一个私密的空间，在外面也能够看见这个庭院，同时与哈得孙河公园的草木也建立了一种视觉上的连接。建筑的斜坡考虑到了南边低层结构和北边、西边高层住宅之间的过渡。

引人注意的非常显眼的屋顶是由一个简单的平行表面组成的，并且带有露台——每个露台都独具特色并且是朝南的。它封闭的外观和建筑的周边形成对比。

建筑的一层和二层主要由不同规模的文化和商业单元构成。

轻微地转变内部墙体的角度，所有的单元就都朝向了哈得孙河，并且向阳。在立面中也会发现墙体上鱼骨的图案。每个公寓都有一个凸出的窗户来获得最壮观的视野。另外，阳台还能使居民和过路人有一个互动。

纽约正迅速地成为一个绿色的适合居住的城市。哈得孙河区和绿色公园的转变，经过不断努力种植的百万棵树，百老汇的步行街和更多的自行车小路的开发，使得整个哥本哈根城成为城市中的绿洲。在西57街我们正在尝试继续把这种转变应用到城市构造的心脏——城市街区的中心。

**West 57th**

The building is an inaugural North American project of BIG. The building's program consists of over 600 residential units of different scales situated on a podium with a cultural and commercial program. The building will strive for LEED Gold Certification.

West 57th introduces an entirely new residential typology to New York City that will add an inviting twist to the Manhattan Skyline.

BIG has taken a step back to survey the rules of the system in Manhattan and is introducing a European typology: the perimeter block. With an efficient layout and a sense of intimacy, the perimeter block meets demands for density and security.

In this project, we have married it to the traditional Manhattan high-rise, creating a unique shape which combines the advantages of both: the compactness of a courtyard building with the airiness and the amazing views of a skyscraper. The form of the building shifts depending on the viewer's vantage point. From

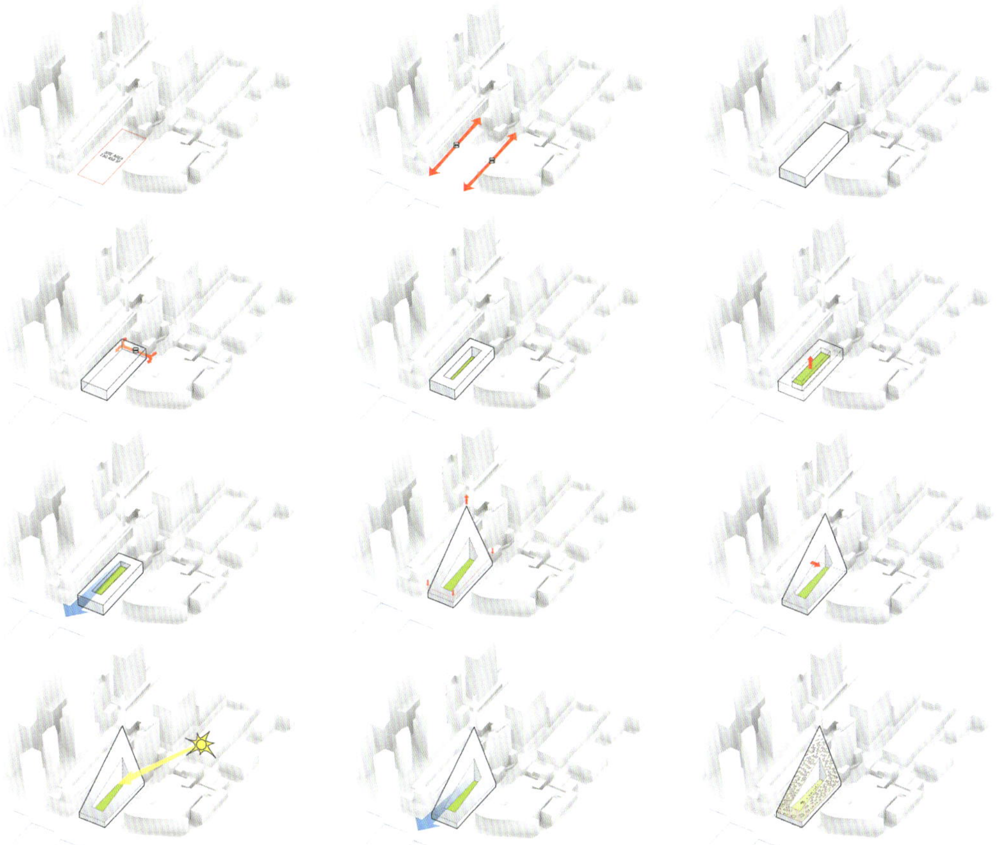

项目名称：West 57th Street
建筑师：BIG-Bjarke Ingels Group
项目主管：Beat Schenk
项目建筑师：Sören Grünert
项目团队：Thomas Christoffersen, Celine Jeanne, Daniel Sundlin, Alessandro Ronfini, Aleksander Tokarz, Alessio Valmori, Alvaro Garcia Mendive, Felicia Guldberg, Gabrielle Nadeau, Ho Kyung Lee, Julian Liang, Julianne Gola, Lucian Racovitan, Marcela Martinez, Maria Nikolova, Minjae Kim, Mitesh Dixit, Nicklas Rasch, Riccardo Mariano, Stanley Lung, Steffan Heath, Thilani Rajarathna, Xu Li
合作者：SLCE Architects(architect of record), Starr Whitehouse Landscape Architects, Thornton Tomasetti(sturctural), Dagher Engineering(MEP), Langan Engineering(civil), Hunter Roberts(construction manager), Philip Habib & Assoc.(transportation), Israel Berger & Assoc. (building envelope), Nancy Packes(marketing), Van Deusen & Assoc.(vertical transportation), Cerami & Assoc.(acoustical), CPP(wind), AKRF(environmental), German Glessner(renderings & animation)
甲方：Durst Fetner Residential
地点：Manhattan, New York, USA
基地面积：80,000m²
获得项目情况：direct commission

the West Side Highway, it appears to be a pyramid; from West 58th, a dramatic glass spire.

By keeping three corners of the block low and lifting the north-east portion of the building, the courtyard opens views towards the Hudson River and brings the low western sun deep into the block. While the courtyard is a private space for residents, it can still be seen from the outside, creating a visual connection to the greenery of the Hudson River Park. The slope of the building allows for a transition in scale between the low-rise structures to the south and the high-rise residential towers to the north and west of the site.

The eye-catching and highly visible roof consists of a simple ruled surface perforated with terraces – each one unique and south-facing. Its closed appearance is in contrast with the building's perimeter.

The building is predominantly residential units of different sizes with cultural and commercial program at the street level and the second floor.

By slightly angling the interior walls, all units are oriented towards the Hudson River and the sun. The fishbone pattern of the walls can also be found in the elevation. Every apartment has a bay window to take advantage of the site's spectacular views. In addition, the balconies encourage interaction between the residents and the passers-by.

New York is rapidly becoming an increasingly green and livable city. The transformation of the Hudson River waterfront and the Highline into green parks, the ongoing effort to plant a million trees, the pedestrianization of Broadway and the creation of more miles of bicycle lanes than the entire city of Copenhagen are all evidence of urban oases appearing all over the city. With West 57th we attempt to continue this transformation into the heart of the city fabric – into the center of a city block. BIG-Bjarke Ingels Group

# 三叶草地

如何在一座已完善的城市中再增加2000所住宅呢?

我们计划在这一地区兴建几个运动场、一个公园以及一条绵延3000m的建筑群。在这一建筑区域内，住宅、托儿所、学校、体育设施、商店以及办公楼等各种市政设施鳞次栉比，就好像中国的长城般蜿蜒环绕着这一地区。三叶草街区这种分化型的高度和序列的设计理念为该街区与周边城市的交通往来提供了很大的便利，同时这种开顶的设计也为居民和游人提供了休闲散步的场所。三叶草街区的设计也正与社会民主党派市长希望在丹麦首都建造新型经济适用房的计划不谋而合。

## Clover Field

How do you fit 2,000 new apartments in the middle of a city that is already complete?

We propose to embrace a vast area of sports fields and a park with a three kilometers long linear building block. A Great Wall of China that winds around the site as it unfolds its mixed use urban program: housing, day care institutions, a school, sports facilities, shops and offices. The design of the differentiated heights and sequence of the Clover Block provides a high level of accessibility to the surrounding city, and the top roof offers promenades for citizens and visitors alike. The Clover Block also seeks to realize the social democratic mayor's plan for new affordable housing in the capital of Denmark.

# 乐高塔

在马歇尔时代，处于战后重建时期的丹麦倾向于建造以预制混凝土为材料的建筑物。因此，现浇混凝土工业几乎绝迹，预制模块与混凝土构件占据了整个建筑产业。如今的丹麦已经俨然成为了一个由乐高积木搭建起来的国家。

若要把停车场、零售商店、办公大厦和住宅等不同种类的密集型建筑集中于一个超密集的狭小区域内，显然普通的外围建筑无法胜任。

我们再一次把流行的尖塔与现代巨石进行对比，试图就此宣告哥本哈根高层建筑的终结。似乎直到20世纪初，建筑师们还一直以设计具有独创性的尖塔来表达对教堂和国王的敬意。但随着功能主义的诞生，我们的想象力似乎已经衰退到只会一再地建造毫无二致的长方形建筑。时至今日，我们还能否发掘并利用功能主义中的有利因素并将其转化为一个创意的平台而非刻板的程式?

## Lego Towers

During the Marshall years when post war Denmark was being rebuilt, the state chose to favor prefabricated concrete over all other forms of construction. As a result the in-situ cast concrete industry has almost vanished, leaving the entire building industry based on prefabricated modules and concrete elements. Contemporary Denmark has become a country entirely made from Lego bricks.

When asked to do a dense mixed use development comprising parking, retail, offices and housing into a single super dense development of a rather small but central site, we knew that the program could not fit in a generic perimeter block.

Once again comparing the popular spires to the modern monoliths, we attempted to diagnose the demise of the Copenhagen high-rise. It appeared that until the early 1900s architects had been capable of designing imaginative towers in the honor of the church or the king. But upon the arrival of functionalism, our imagination seemed to have been reduced to the repetition of identical rectangular slabs. Could we today harness all the good intentions of functionalism and turn it into a platform of the creative process, rather than its straight jacket?

# 环形城市

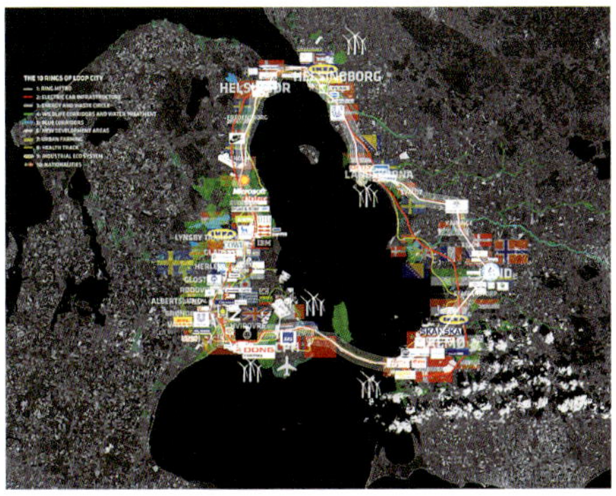

哥本哈根城郊的工业区即将成为下一个城市开发区域。一条新的轻轨线路正在拟建中，它将连接整个城市中心区内占地共11km²的20个开发区域。我们期望这条轻轨线路可以成为高峰期的主要交通手段。

我们期望通过延长这条轻轨线路，使之环绕奥里桑德地区并连接周围同类的开发区域，从而满足未来50年瑞典、与丹麦接壤地区的发展需求。环形城市连接着无中心大都市区内高度差异化的城市节点、大学和工作场所，这正是1947年“手指规划”中所提出的城郊与市中心的连接处。这一区域在规模上可与旧金山海湾地区相媲美，拥有着足以应对未来50年发展的巨大空间。环形城市定将成为该地区可持续发展、密集型、超级娱乐开发项目的一个典范。

## Loop City

The industrial areas in the Copenhagen suburbs are next in line for urban development. A new light rail is planned to interconnect 20 development zones with a total area of 11 km², the size of the entire inner city. We are proposing to turn the light rail line into a spine of dense urbanity with a series of peaks at the stations.

We propose to upgrade the planned light rail by extending it to form a regional ring around Oresund connecting similar development areas, and creating a new 50 year development perspective for a cross border region between Sweden and Denmark. Where the Finger Plan from 1947 was about connectivity from suburb to center the Loop City is linking a string of highly differentiated urban nodes, universities and working spaces in a center-less metropolitan region around a blue void. This ring, comparable in size to the San Francisco Bay area, has enough development areas to contain the growth of the region the next 50 years, and could become a model for a both sustainable, dense and super recreational development of the region.

# 丹麦馆

2010年上海世博会上，丹麦馆为游客提供了一个可以亲身体会丹麦城市生活中最精彩部分的机会。在交往与互动中，游客得以体验到哥本哈根最具吸引力的事物——城市自行车、港口游泳池、运动场设施、屋顶花园的野餐，并有幸见到了安徒生的小美人鱼。

游客可以在两种速率中感受此次展览会，既可以悠闲地漫步于馆内大饱眼福，也可以骑上动感的自行车来感受这个城市和城市生活的快节奏。

丹麦馆设计为一条可供城市自行车与行人通行的交通环线。就如丹麦城一样，我们最好通过步行或骑自行车来感受丹麦馆的风情。丹麦馆顶层排列着300多辆自行车，免费供游客使用，为大家提供了一个感受丹麦城市生活方式的绝佳机会。

## Danish Pavilion

The Danish Pavilion at Shanghai Expo 2010 gave visitors the opportunity to try some of the best aspects of Danish city life themselves. Through interaction, the visitors were able to actually experience some of Copenhagen's best attractions – the city bike, the harbor bath, playground settings, a picnic on the roof garden and the opportunity to see the authentic H.C Andersen's Little Mermaid.

The exhibition was experienced in two speeds, as a calm stroll with time to absorb the surroundings and as a dynamic bicycle trip, where the city and city life rush past.

The Pavilion is designed as a traffic loop created by the motion of city bikes and pedestrians tied in a knot. Like a Danish city, the Danish Pavilion was best experienced on foot and by bike. Over 300 free city bikes located upon the roofscape, offer the visitors a chance to experience the Danish urban lifestyle which includes biking everywhere.

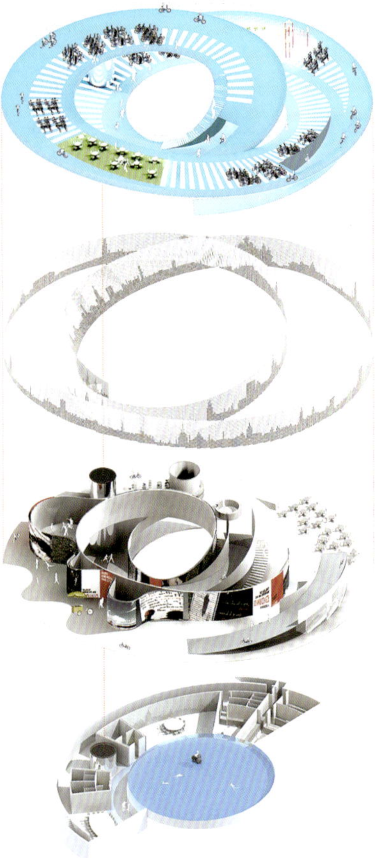

courtesy of the architect(©Iwan Baan)

# 山阶住宅

如何将郊区院落的美妙之处与城市生活的紧张节奏完美地结合起来？

山阶住宅是VM住宅的第二代产物。山阶住宅这一项目三分之二的面积用于泊车，三分之一用于居住。停车区需要与街道相连通，住宅需要充足的阳光、新鲜的空气和良好的视野。因而在山阶住宅中，所有的公寓都有朝阳的屋顶花园、宽阔的视野以及十层楼高的停车区。山阶住宅为居民提供了同时享受两种生活优点的机会：既靠近哥本哈根市中心繁华的城市生活，又能领略到乡下宁静的田园风光。

山阶住宅中80户居民将成为奥瑞斯塔丹市首批能够直接在家门口停车的市民。山阶住宅庞大的停车区提供了480个停车位以及沿着内墙设置的坡形电梯。山阶住宅的北侧和西侧覆盖着穿孔铝板，这些孔洞可以为停车区提供空气和阳光。

## The Mountain

How do you combine the splendors of the suburban backyard with the social intensity of urban density?

The Mountain is the 2nd generation of the VM Houses. The program, however, is 2/3 parking and 1/3 living. The parking area needs to be connected to the street, and the homes require sunlight, fresh air and views, thus all apartments have roof gardens facing the sun, amazing views and parking on the 10th floor. The Mountain appears as a suburban neighborhood of garden homes flowing over a 10-story building – suburban living with urban density.

The residents of the 80 apartments are the first in Orestaden to have the possibility of parking directly outside their homes. The gigantic parking area contains 480 parking spots and a sloping elevator that moves along the mountain's inner walls. The north and west facades are covered by perforated aluminum plates, which let in air and light to the parking area.

courtesy of the architect(©Jens Lindhe)

courtesy of the architect(©Jakob Boserup)

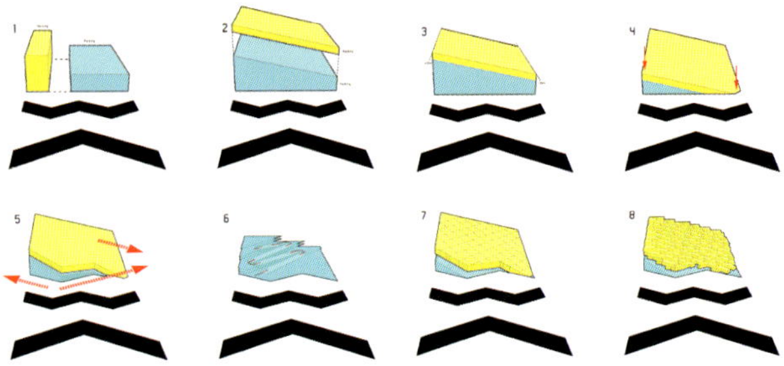

# 8字之家

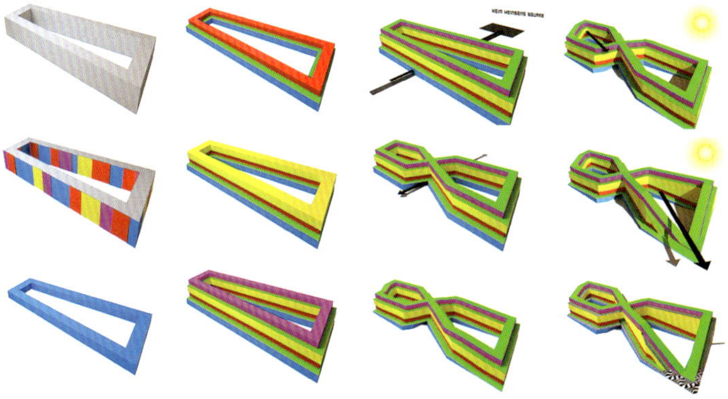

8字之家秉承着在大环境中注重细节的建筑理念。8 字之家是一个面积达60 000m²的集合住宅，它是一个居住和商务并存的空间：既有乡村般的宁静，也有大都市的活力。这个弓形建筑创造了两个截然不同的空间，这两个空间被处于弓形中心处的一个500m²的公共设施隔离开来。建筑师并没有根据居住和商用功能把该建筑分成独立的体块，而是将各种功能布置在水平的方向上：居住公寓布置在上面，商用楼层在整栋建筑的底部展开。因此，不同的水平分层实现了自身的品质：公寓的视野开阔，光线充足，空气新鲜，而出租的办公区与街区生活融为一体。

## 8 House

8 House is where you will find the attention to detail embedded in a larger context. Here, closeness thrives in the 60,000m² building. This is where the tranquility of suburban life goes hand in hand with the energy of a big city, where business and housing co-exist. The bow-shaped building creates two distinct spaces, separated by the center of the bow which hosts the communal facilities of 500m². Instead of dividing the different functions of the building for both habitation and trades into separate blocks, the various functions have been spread out horizontally. The apartments are placed at the top while the commercial program unfolds at the base of the building. As a result, the different horizontal layers have achieved a quality of their own: the apartments benefit from the view, sunlight and fresh air, while the office leases merge with life on the street.

courtesy of the architect (©Ty Stange)

courtesy of the architect (©Dragor Luftfoto)

# VM住宅

如果人各不相同，那么为什么所有的房间都是一样的?

VM住宅由一个"V"形的和一个"M"形的居住用的体块组成。该建筑采用这样的体块组合方式以满足居住者对于阳光、私密性和视线的需要。通过推动中央的平板，在消除邻居面对面的缺点的同时，还能满足居住者观看周围这片大而且开阔的场地的需要。所有的房间都有双层高度的朝北空间和朝南的全景视野。建筑中心的走廊很短并且能从两端接收到阳光，像弹孔穿透建筑。VM住宅为居住者提供了80多种设计灵活的不同户型，满足了当代生活的个性化需求。

## VM Houses

If people are different then why are all apartments alike?

The VM Houses are two residential blocks formed as the letters V and M. The blocks are formed as such to allow for daylight, privacy and views. The vis-a-vis with the neighbor is eliminated by pushing the slab in its center, ensuring diagonal views to the vast and open, surrounding fields. All apartments have a double-height space to the north and wide panoramic views to the south. The central corridors are short and receive light from both ends, like bullet holes penetrating the building. The VM Houses offer more than 80 different apartment types that are programmatically flexible and open to the individual needs of contemporary life.

courtesy of the architect (©Johan Fowelin)

courtesy of the architect (©Nils Lund)

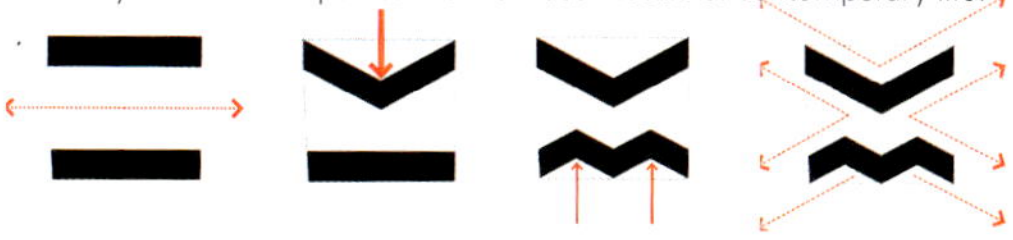

courtesy of the architect (©Dragor Luftfoto)

## >>56

### David Jameson Architect

David Jameson graduated from Virginia Polytechnic Institute and State University in 1991 and found David Jameson Architect in 2001. Was a member of AIA Northern Virginia Board of Directors and Small Firms Committee Chair from 2004 to 2009. Has worked to communicate the value of design excellence to the public. Volunteered as a class leader for the AIA Architects in the school program to advance the knowledge of elementary school children about architecture and design. Was an award recipient of 2011 AIA SPP Honor Award for Architecture for Tea House.

## >>74

### Noa Biran + Roy Talmon

Their design process combines architecture - photography - dancing - video art with wider fields of culture and society. Dealing with architecture projects as well as artistic projects and installation, Noa Biran[left] and Roy Talmon[right] try to stretch the ground of their creation, including in it their political - cultural and sociological points of view, as well as the environmental and ecological beliefs. Also working cultural and communal projects, they work together independently on private houses, interior design and architecture installations. Lately, presented a project, at the Bat Yam international Biennale of landscape urbanism, dealing with the hazards happened because of construction fences.

## >>66

### Ravnikar Potokar Architectural Office

Robert Potokar graduated from the Faculty of Architecture in Ljubljana in 1990. He is a member of Chamber of Architecture and Spatial Planning in Slovenia. Since 1999, he has been a managing editor of the Piranesi Review. He was a recipient of awards of 2008 Slovenian Chamber of Architects award in Ljubljana.

## >>92

### ZigZag Arquitectura

Was founded in Madrid in 2005 by Bernardo Angelini[left] and David Casino[right]. Both architects studied in the ETSAM. They had previously worked for prestigious firms like FOA, Nomad and Ensamble Studio. They are currently working as professors at the UPSAM and the ETSAM, combining their teaching activity with their research work in the studio. They

## >>14

**e|348 Arquitectura**

s an architecture practice based in Póvoa de Varzim. Its main goal is to develop a product capable of increasing human habitat's qualification, with efficiency and innovation. Singularity, comfort, efficiency, economy, responsibility, aesthetics, durability and flexibility, define a few qualities that become part of every project. Each project searches solutions which, n an integrated way, combine clients' expectations with responsible answers, ethically and aesthetically.

## >>32

**Architecture H**

Was founded in 1990 by HyunJai Lee[right]. HyunJai Lee was born in 1947, DaeGu, and studied Fine Arts at an arts college. Was a leading artist in experimental art movement in the 1970's and 1980's. JiWon Lee[left] was born in 1976. She graduated from the College of music, Seoul National University. She is a composer as well as an architect.

## >>70

**Petra Gipp Arkitektur**

Petra Gipp was born in Stockholm in 1967. Graduated from the Royal Danish Academy of Fine Arts in 1999. Her projects were introduced to several books and magazines such as New Small Spaces(Francesc Zamora Mola, 2008), Arkitektur(2010.4), Dagens Nyheter(2010).

## >>26

**Shinichi Ogawa & Associates / urbanist architect**

Shinichi Ogawa was born in Yamaguchi, Japan in 1955. Received Bachelor of Fine Art from Nihon University, College of Arts in 1978. Established Shinichi Ogawa & Associates / urbanist architect in 1986. Was a professor at the Kinki University, Faculty of Engineering in 2008.

## >>62

**A1 Architects**

Was set up by Lenka Křemenová[left] and David Maštálka[right] in Prague, Czech republic. I. operated in the wide range of architectural designs and graphic projects. Their work is characterized as the interdisciplinary communication and the quest for unique solutions interknitting different artistic points of view.

## >>50

**5468796 Architecture**

s a Winnipeg architecture collaborative with expansive interests engaging n all aspects of design including visioning, branding, architecture, object design, detailing and engineering systems. The office was formed in 2007 and continually seeks to challenge convention in a multitude of scales. The studio constantly dissects design and culture through a series of ongoing dialogues and explorations. Creativity is paramount within the studio – not as a manifestation of individual preference, but rather as a direct response o client, context and program.

## >>158

**BIG – Bjarke Ingels Group**

Was founded in 2005 by Bjarke Ingels. Is an architectural office currently involved in a large number of projects throughout Europe, Asia and North America. BIG's architecture emerges out of a careful analysis of how contemporary life constantly evolves and changes, not least due to the influence of multicultural exchange, global economic flows and communication technologies that together require new ways of architectural and urban organization.

## >>38

**BNKR Arquitectura**

Is an architecture, urbanism and research office founded by Esteban Suárez[left] in 2005, Mexico City and partnered by his brother Sebastián Suárez[right]. In their short career they have been able to experience experiment architecture in the broad scale: from small iconic chapels for private clients to a master plan for an entire city. Bunker's unconventional approach to architecture has continuously generated public debate with projects such as a three-kilometer habitable bridge that unites the bay of Acapulco and an inverted skyscraper 300 meters deep in the main square of the historic center of Mexico City.

## >>108

**Frédéric Schlachet Architecte**

Was founded in 2001. Frédéric Schlachet studied in Versailles Architecture School. In 2010, he taught architectural projects in ENSA Versailles. Since 2004, he has been working on various projects in Paris, such as social housing buildings, primary schools and airports.

## >>118

**Magén Arquitectos**

Is an architectural practice founded by Jaime Magén and Francisco J. Magén based in Zaragoza since 2002. Is interested in Architecture as a both technical and cultural fact, deeply focused on unique investigation and development of each project. Jaime Magén studied Architecture at ETSA and graduated from Navarra University in 1999. Established his own firm in 2001. In 2004, the firm changed the name to Magén Arquitectos. Francisco J. Magén studied at the same university and graduated in 2004. Became a partner at Magén Arquitectos since 2004.

## >>80

**Ivo Pavlik**

Was born in 1979 in Dukovany, Czech Republic. Graduated from Technical University of Liberec, Faculty of Art and Architecture in 2010. In the same year, he founded architectural office, Hand Mind Makers, with Lucie Chytilova, Dana Novakova, Ilona Zitnakova, Ondrej Vosatka.

## >>10

**Andrea Tognon**

founded an architecture and communication design research laboratory in 2002 with the aim of understanding and developing strategies of communication and branding, modeling and transforming spaces, defining identity in the dynamic process of the complexity of urban situations. The office combines practice with research in a multi-disciplinary approach while drawing upon a network of creative, technical and engineering firms with whom integrate different aspects of the design process and construction. Is based in Milan and working across the globe.

## >>152

**Caramel Architekten**

Günter Katherl[middle], Martin Haller[left] and Ulrich Aspetsberger[right] have been working together since 2000 and with the help of their team have already completed a number of building projects of varying dimensions. Their most recent commissions have resulted from their continual and successful participation in national and international competitions. In addition to the realization of the large-scale projects like the science park Johannes Kepler University in Linz, office and central garage in Ansfelden, the trio also devotes itself in particular to design studies and unconventional single-family dwelling projects.

## >>144

**C. F. Møller Architects**

Is one of Scandinavia's oldest and largest architectural practices. Was founded in 1924 by prof. C.F. Møller. Today is a partnership owned and managed by nine partners, Julian Weyer, Mads Mandrup, Tom Danielsen, Anna Maria Indrio, Klavs Hyttel, Lars Kirkegaard, Lone Wiggers, Mads Møller and Klaus Toustrup. Over the years, has won a large number of national and international competitions. Also, their work has been exhibited all over the world as well as published in books and leading professional journals.

## >>126

**Garduño Arquitectos**

In 1990, Juan Garduño graduated with honors from Universidad Anáhuac. In 1994, he founded Garduño Arquitectos, an architectural group envisioned as a workshop. It was conceived as a professional, innovative and environmentally-responsible firm that pays close attention to every detail of the project, taking on one item at a time. Their works are the recipient of continuous recognition and has been published

## >>134

**O'Donnell + Tuomey**

Was established by Sheila O'Donnell and John Tuomey in 1988. Has been involved with urban design, cultural and educational buildings, houses and housing projects in Ireland, the Netherlands and the UK. They were members of Group 91 and winners of more than 50 national and international awards for their work across the past twenty years. Published a monograph O'Donnell + Tuomey Selected Works in 2007.

著作合同登记06-2011年第236号



**图书在版编目(CIP)数据**

小型建筑：中文、英文/韩国C3出版公社编；王单单，李群译．—大连：大连理工大学出版社，2011.10
（C3建筑立场系列丛书；5）
ISBN 978-7-5611-6579-9

Ⅰ.①小… Ⅱ.①韩… ②王… ③李… Ⅲ.①建筑设计—汉、英 Ⅳ.①TU2
中国版本图书馆CIP数据核字（2011）第212662号

出版发行：大连理工大学出版社
（地址：大连市软件园路 80 号　　邮编：116023）
印　　刷：利丰雅高印刷（深圳）有限公司
幅面尺寸：225mm×300mm
印　　张：12.25
出版时间：2011 年 10 月第 1 版
印刷时间：2011 年 10 月第 1 次印刷
出 版 人：金英伟
统　　筹：房　磊
责任编辑：杨　丹
封面设计：王志峰
责任校对：张媛媛

书　　号：ISBN 978-7-5611-6579-9
定　　价：150.00 元

发　行：0411-84708842
传　真：0411-84701466
E-mail: a_detail@dutp.cn
URL: http://www.dutp.cn